SAPIENTS
THE NEXT EVOLUTION OF CONSCIOUSNESS,
COGNITION AND EMOTIONAL INTELLIGENCE
J.L. MEE

SAPIENTS

The Next Evolution of Consciousness, Cognition and Emotional Intelligence

PROFESSIONAL EDITION

ISBN 978-1-892654-5-02

Published in the United States of America

Arcturus Press
arcturuspress.com

CONTENTS

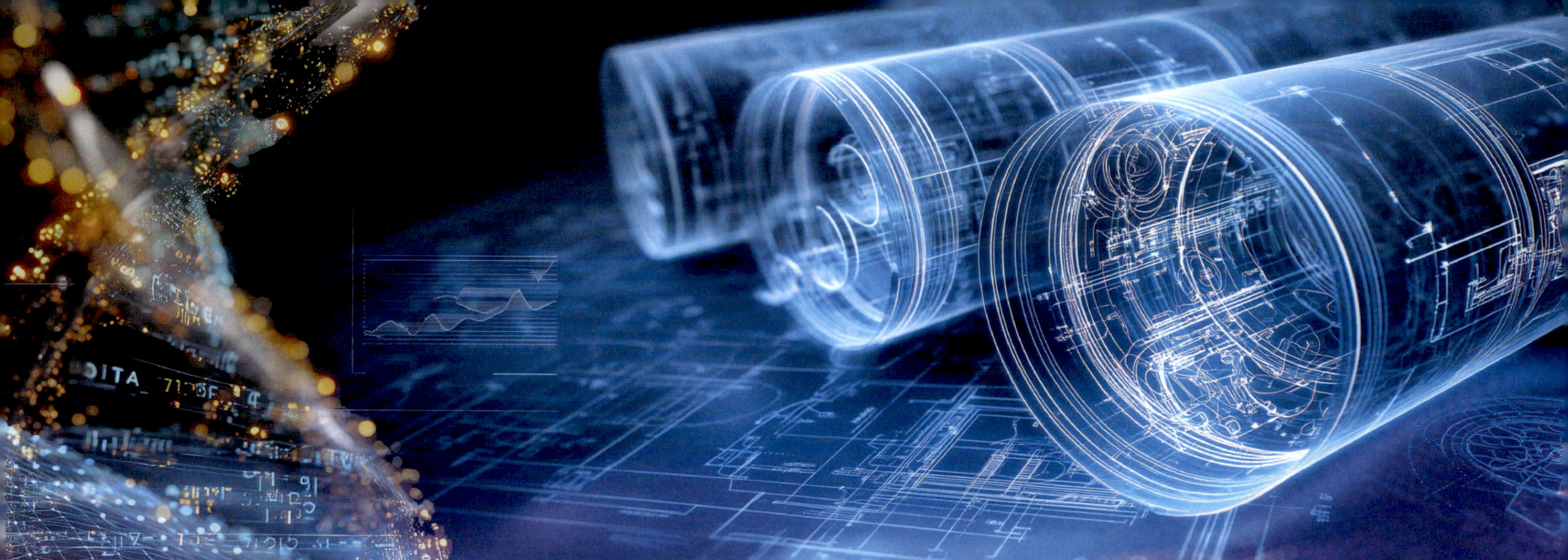

Preface

This book presents the design blueprints, foundational science and enabling technology for evolving homo sapiens to a new form.

The Sapient will possess advanced cognitive abilities, including self-awareness, emotional intelligence and metacognition.

Sapients will demonstrate an elevated level of consciousness, insight and wisdom. They will be capable of self-directed learning and adaptive decision-making.

These goals will be realized through the application of a new science of awareness which provides the technology for genetically modifying human neurology to optimize cognitive performance.

Key Concepts

Cognitive Performance

Genetically improving cognition in adults is becoming feasible.

Treated individuals will enjoy a wide range of personal and professional benefits.

Best practices have been identified for delivering cognitive upgrades.

Scientific Discovery

New research has revealed that conscious awareness varies inversely with the brain's electric current power level.

This discovery is broadly supported in scientific literature. It illuminates a pathway to genetic cognitive enhancement.

Cognitive Engineering

The brain's power level can be adjusted by genetically modifying the electrical properties of neurons.

A minor change to a small percent of neurons can attenuate the brain's electric field and expand cognitive capacity.

Gene therapies for temporarily modifying neurons have demonstrated safety and efficacy in preclinical studies.

Responsible Deployment

These discoveries have far-reaching implications for our world.

Cognitively-enhanced humans can have a beneficial impact on society, but misuse of the technology can exacerbate social inequalities.

A framework with ethical standards for responsibly deploying the technology has been established.

Homo Sapiens Limitations

Humanity has arrived at an evolutionary crossroad. Despite having extraordinary cognitive potential, we use only a fraction of our abilities.

Modern humans are plagued by compulsive thinking, limited self-awareness and emotional imbalance; all issues exacerbated by the blistering pace and demands of contemporary life.

Traditional methods of self-improvement such as meditation attempt to address these challenges, but do so only incrementally and inefficiently.

Meanwhile, critical global challenges demand solutions from individuals operating at their highest cognitive and emotional potential.

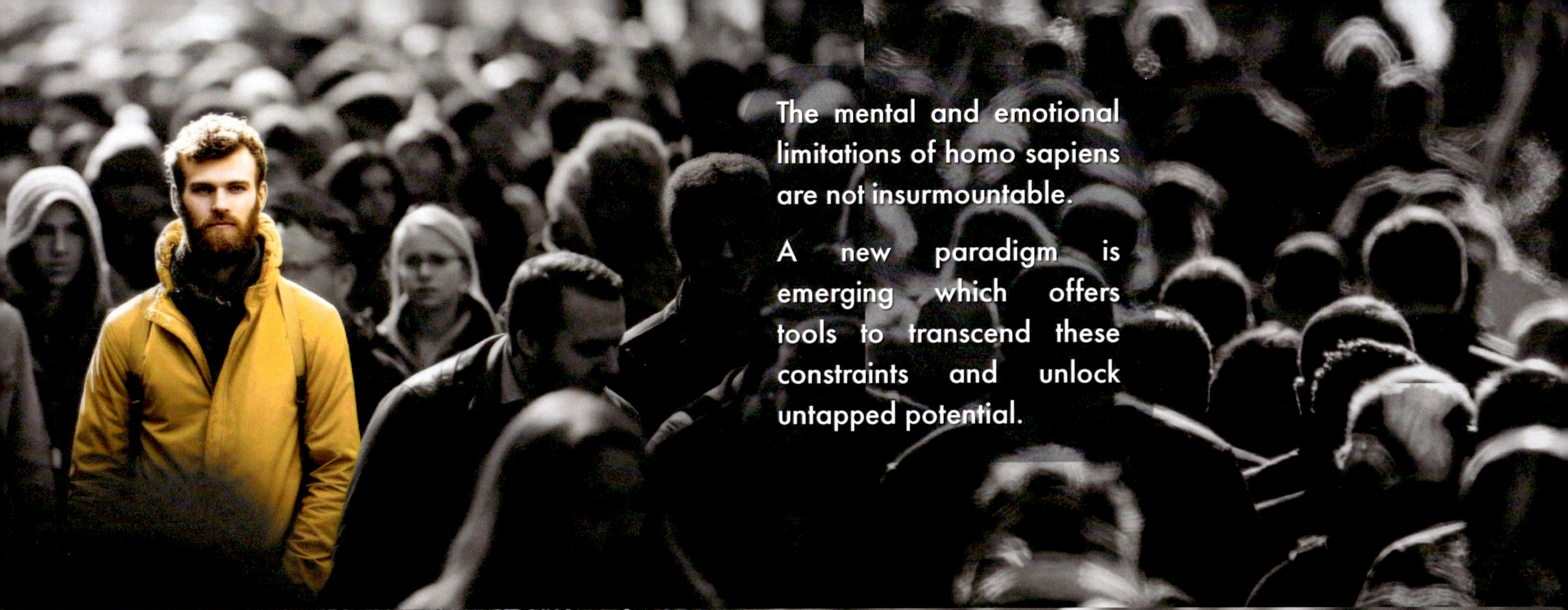

The mental and emotional limitations of homo sapiens are not insurmountable.

A new paradigm is emerging which offers tools to transcend these constraints and unlock untapped potential.

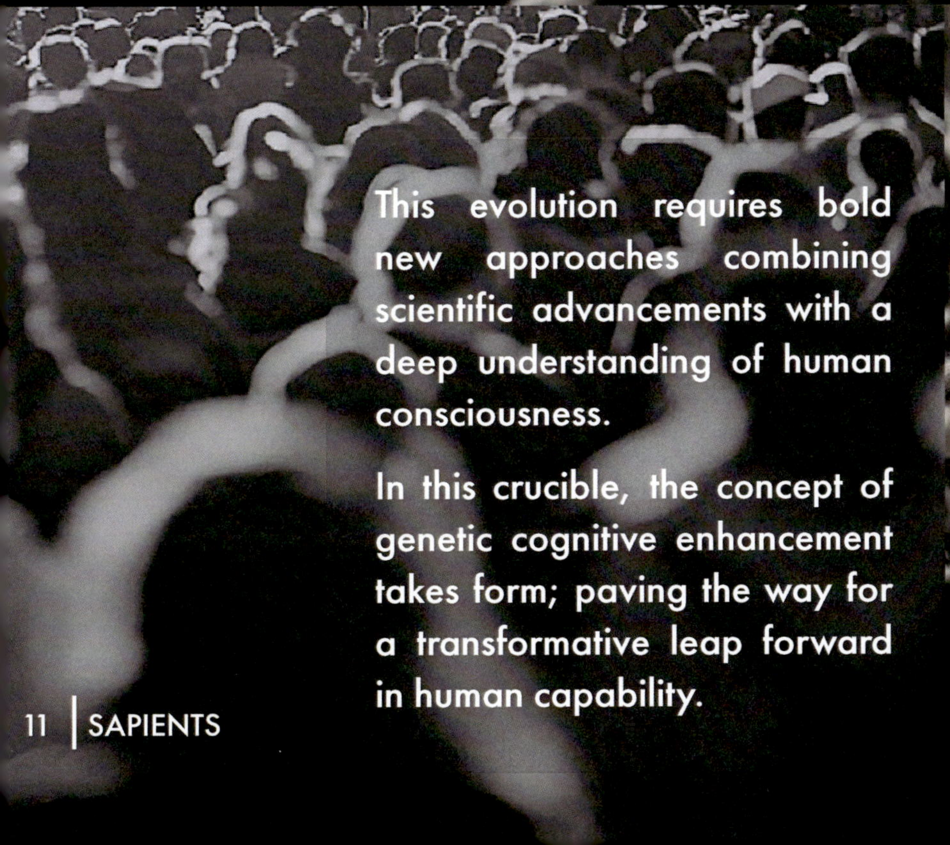

This evolution requires bold new approaches combining scientific advancements with a deep understanding of human consciousness.

In this crucible, the concept of genetic cognitive enhancement takes form; paving the way for a transformative leap forward in human capability.

Becoming Sapient

The Sapient transformation will modify the architecture of consciousness itself; changing how individuals engage with the world.

Mindfulness will become second nature, allowing people to navigate life with poise and authenticity. Values will be guided by insight, vision, and inner alignment.

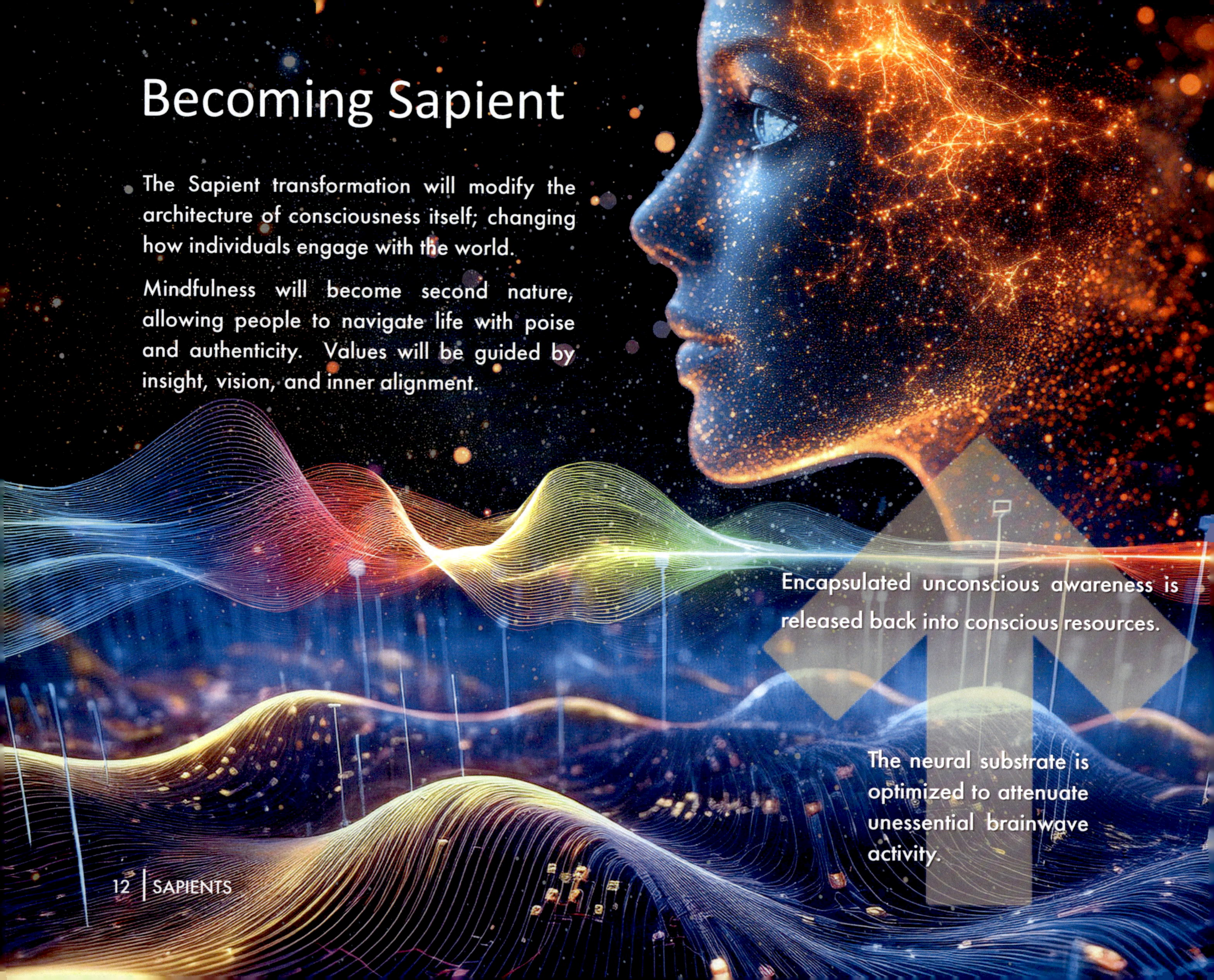

Encapsulated unconscious awareness is released back into conscious resources.

The neural substrate is optimized to attenuate unessential brainwave activity.

The Master Control Switch

The formulas in this book define the architecture for genetic neuroengineering techniques for promoting cognitive flexibility and enhanced awareness.

Research within this framework has revealed the master control switch for consciousness in the human genome: Gene HTR2A on chromosome 13. Modifying this gene can dramatically improve attention, memory and emotional resilience.

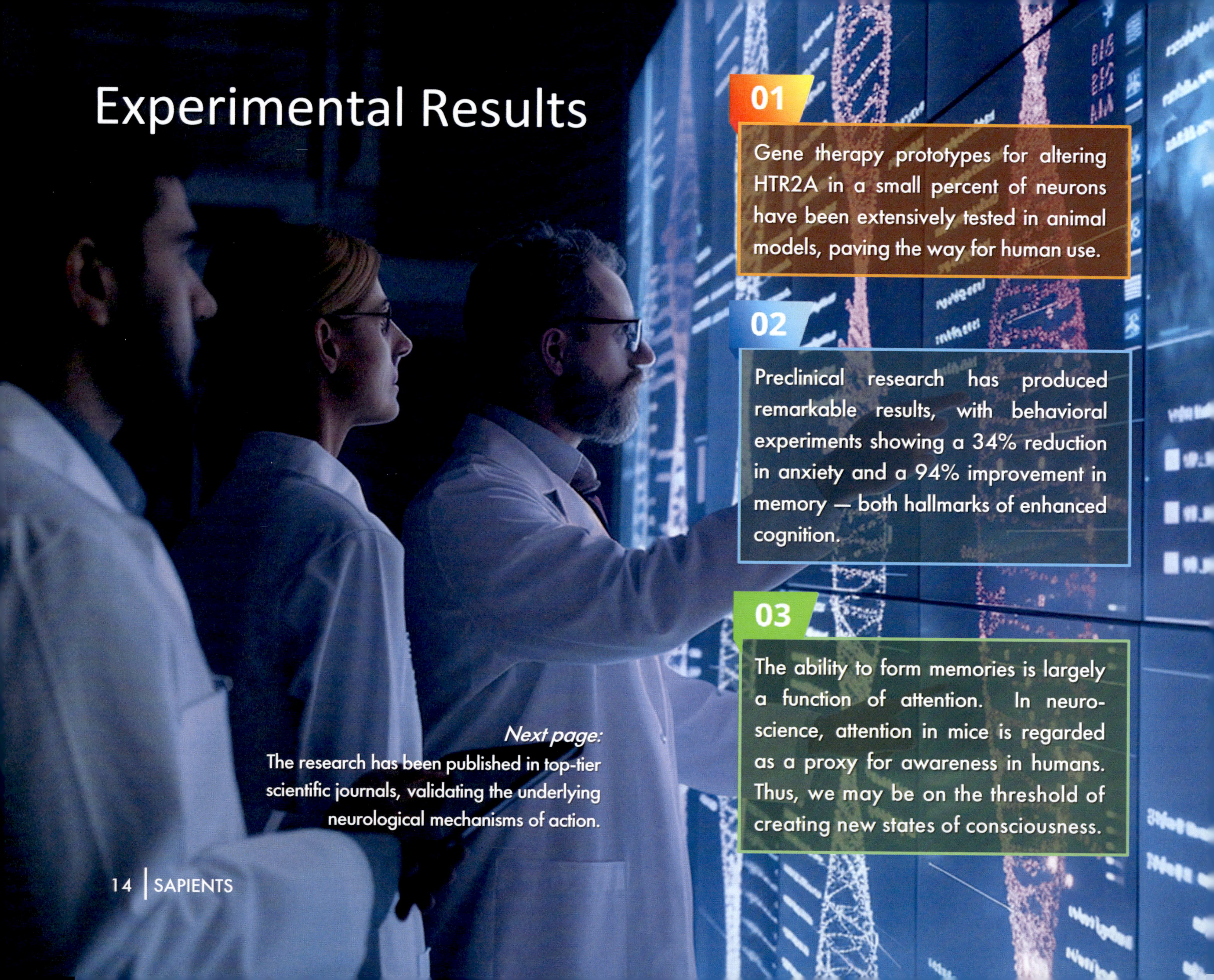

Experimental Results

01

Gene therapy prototypes for altering HTR2A in a small percent of neurons have been extensively tested in animal models, paving the way for human use.

02

Preclinical research has produced remarkable results, with behavioral experiments showing a 34% reduction in anxiety and a 94% improvement in memory — both hallmarks of enhanced cognition.

03

The ability to form memories is largely a function of attention. In neuroscience, attention in mice is regarded as a proxy for awareness in humans. Thus, we may be on the threshold of creating new states of consciousness.

Next page:
The research has been published in top-tier scientific journals, validating the underlying neurological mechanisms of action.

ga Pro DATA & APIS EVENTS PREMARKET ADVERTISE

ZINGA Our Services News Markets Options Ratings Ideas Money Crypto

SPY	QQQ	BTC/USD	DIA	GLD	TLT
521.23 ▼ 1.72%	553.05 ▼ 2.12%	112950.00 ▲ 0.3776%	436.60 ▼ 1.07%	309.74 ▲ 2.24%	87.79 ▲ 1%

June 22, 2023 1:50 PM 2 min read

Cognigenics' Intranasal CRISPR Delivery Bypasses Blood-Brain Barrier; Potential Breakthrough Therapeutic

EurekAlert! AAAS

NEWS RELEASES MULTIMEDIA MEETINGS

RELEASE 5-SEP-2024

...w RNA treatment shows promise for ...hancing memory and reducing anxiety

...ative shRNA therapy targeting serotonin receptor improves cognitive function in animal ...s, offering potential for treating mild cognitive impairment

Genomic Press

Genomic Psychiatry

Advancing science from genes to society

SCIENTIFIC INQUIRER

Everyday Science Left Brain/Right Brain Sports Features About Contact

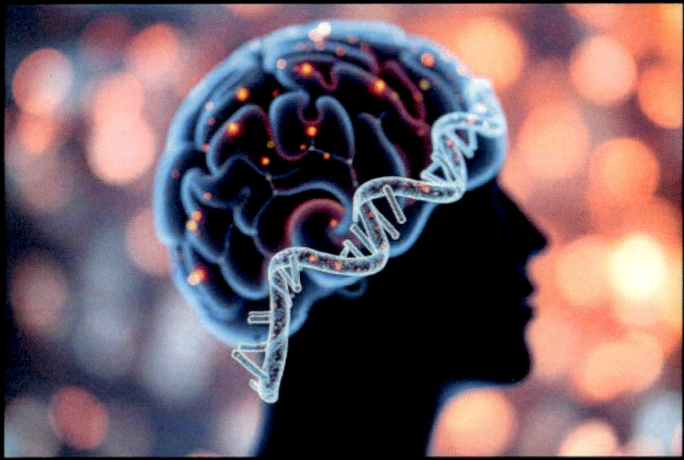

Everyday Science

New RNA treatment shows promise for enhancing memory and reducing anxiety

Cognigenics' research on COG-201 in Genomic Psychiatry shows promising effects on memory and anxiety in animal models through RNA therapy

scientificinquirer

...FORD ACADEMIC Journals Books

PNAS NEXUS

Advance articles Subject ▾ Submit ▾ Alerts About ▾ PNAS Nexus ▾

yahoo!finance

Cognigenics Announces Groundbreaking Publication on shRNA Treatment for 5-HT2A Receptor Knockdown

Neuroscience News.com

RNA Therapy Improves Memory and Reduces Anxiety

Featured Genetics Neuroscience · September 5, 2024

JOURNAL ARTICLE

Genetic modulation of the *HTR2A* gene reduces anxiety-related behavior in mice

Troy T Rohn, Dean Radin, Tracy Brandmeyer, Barry J Linder, Emile Andriambeloson, Stéphanie Wagner, James Kehler, Ana Vasileva, Huaien Wang, John L Mee ... Show

The Cognitive Age

The research indicates a large-scale breakthrough with the potential for translating into widespread human use.

When clinical trials produce favorable results, humanity will enter a new phase in its evolution: The Cognitive Age

Societies, groups and nations which embrace this disruptive new technology can harness its vast potential to drive exponential growth in cognitive capital to help meet the challenges facing the world today.

1.1: Sapience

Genetic upgrades can potentially evoke a myriad of desirable cognitive and emotional traits:

Metacognition – 5

Mood – 4

Clarity – 3

Presence – 2

Perception – 1

Preparatory levels will equip ndidates to manage the spectrum of Sapience awareness characteristics.

1.2: Personal Value

For Individuals

The Sapient transformation can augment individual abilities to serve each person's unique needs and goals.

Personal Growth: Achieving higher consciousness and greater clarity of purpose, accelerating personal evolution.

- Learning: Building new skills more quickly and improve academic performance.
- Meditation: Experiencing the benefits of meditation effortlessly, fostering inner peace, mental clarity and mindfulness.
- Empowerment: Gaining greater sovereignty, expand perspective on life and increase freedom of choice.
- Emotional Well-being: Enhancing love, acceptance, tolerance and understanding.
- Flexibility: Cultivating openness to change and adaptability in life with increased self-empowerment and confidence.

For Families and Relationships

The technology also promises to strengthen relationships and family dynamics:

- Compassion: Increasing understanding, love and acceptance among family members.

- Harmony: Fostering a harmonious and peaceful home environment, promoting family unity and resilience.

- Emotional Intelligence: Enhancing emotional balance and relationship harmony for spouses, deepening love and connection.

- Relationship Depth: Building deeper and more rewarding relationships through improved communication and mutual trust.

- Progeny: Allowing children to form models of the world based on adults who are living at optimum capacity; positively influencing their epigenetic development.

1.3: Professional Value

In a rapidly evolving economy shaped by AI and automation, human-value skills that cannot be digitized – such as imagination, creativity, and intuition – are becoming increasingly valuable.

Thriving in this landscape will require innovation, energy, and focus to stay competitive. Genetic enhancement can release a person's latent natural abilities, providing the mental acuity, resilience and cognitive flexibility to tackle complex challenges and drive their ventures forward.

Cognitive upgrades will increase a person's capacity for creative innovation and learning new competencies; delivering a mental boost for mastering key new skill sets essential for success in the AI economy.

Greater ability to develop creative solutions to business problems will yield advantages in the workplace. Hence, the upgrades will have an economic value.

1.4: Sapient Life

With heightened mental acuity, professional endeavors thrive. Sapients will excel in leadership, problem-solving, and innovation. Tasks that once seemed daunting will be approached with confidence, creativity and focus.

Sapients will cultivate deeper, more meaningful personal and professional relationships. Their inner clarity will guide decisions with precision and purpose, while emotional intelligence brings composure, optimism and insight to human interactions.

1.5: Becoming Supernormal

Becoming a Sapient is more than a genetic enhancement — it is a psychological and philosophical metamorphosis.

Structured learning experiences will align each candidate's mindset with their upcoming cognitive upgrade. This process includes education on life management skills and the science behind the transformation, as well as counseling to ensure emotional readiness.

2.1 Delivery

A central campus of higher learning will serve as the nucleus of a holistic cognitive ecosystem.

Interconnected scientific, business and academic organizations will develop world-class best practices for excellence in genetic cognitive engineering technology and education.

The college will function as nexus for scientific researc and responsible governanc developing ethical framework and benchmarks in tander with advancing the science c cognitive enhancement.

2.2 Enhancement

Cognitive gene therapies will be delivered along with comprehensive education and counseling programs to ensure candidates gain the full advantage of their upgrades.

A Cognitive College will offer a mental software upgrade to maximize user benefits from enhanced neural hardware.

All services will be performed on a central campus to maintain uniform excellence in quality and support.

Cognitive College applicants will complete a series of interviews and psychological tests to verify each person's suitability for gene therapy.

The college will accept candidates with common mental health issues such as stress, anxiety and depression; but not serious conditions, such as psychiatric disorders, which could interfere with a candidate's progress. Applicants who successfully complete admissions testing will be accepted for enrollment.

SAPIENT LEVELS

Level 400
SAPIENT II:
Level 400 selectively reconfigures the Sapient's neuronal genotype to create a new subspecies of homo sapiens with greater presence, clarity and acuity.

Level 300 – SAPIENT I:
Level 300 modifies neural networks to amplify desirable cognitive and emotional traits using temporary RNA therapies.

Level 200 – PRIME:
Level 200 equips Sapient candidates with the knowledge, skills and mindset to successfully navigate their cognitive upgrade journey.

Level 100 – WHOLE:
Level 100 fortifies neural networks to increase resiliency and reduce reactivity; raising self-determinism, confidence and emotional intelligence.

Level 100 – Whole

Balance and Resilience

Mental health therapy is a critical precursor to cognitive enhancement, clearing psychological obstacles and aligning individual goals.

Level 100 strengthens neural function to minimize stress responses, fostering mental stability, adaptability and receptivity to transformation—key elements for cognitive advancement

Embracing boldness, love, and new experiences becomes easier—leading to greater success and fulfillment.

Ultimately, building neurological resilience and anti-fragility fosters the positive behaviors and mindsets which pave the way for sustained achievement, personal growth, and enriched cognitive development.

Level 200 – Prime

Ready for Upgrade

Sapient candidates will attend a structured education and counseling program to prepare them for genetic enhancement and equip them to achieve their personal and cognitive goals.

The Prime program aligns students for a profound life change by cultivating a growth mindset and providing actionable strategies for leveraging the enhancements for maximum benefit.

The curriculum will help students develop the requisite skills for managing living in higher states of awareness and cognitive capacity, and for translating their newfound abilities into meaningful personal growth.

It also familiarizes students with the cognitive upgrade process, its scientific foundations, and its administration protocols. Graduates will gain the expertise, abilities, and perspective to confidently embark on their cognitive enhancement journey.

PRIME PROGRAM :

Part 1 – Foundation: Inner Alignment: Students are prepared for dynamic transformation by re-examining and reframing their openness to change, model of the world, perceptual filters, values, beliefs, attitudes, emotions, expectations, assumptions, memories, decisions, perceptions, thoughts and perspectives.

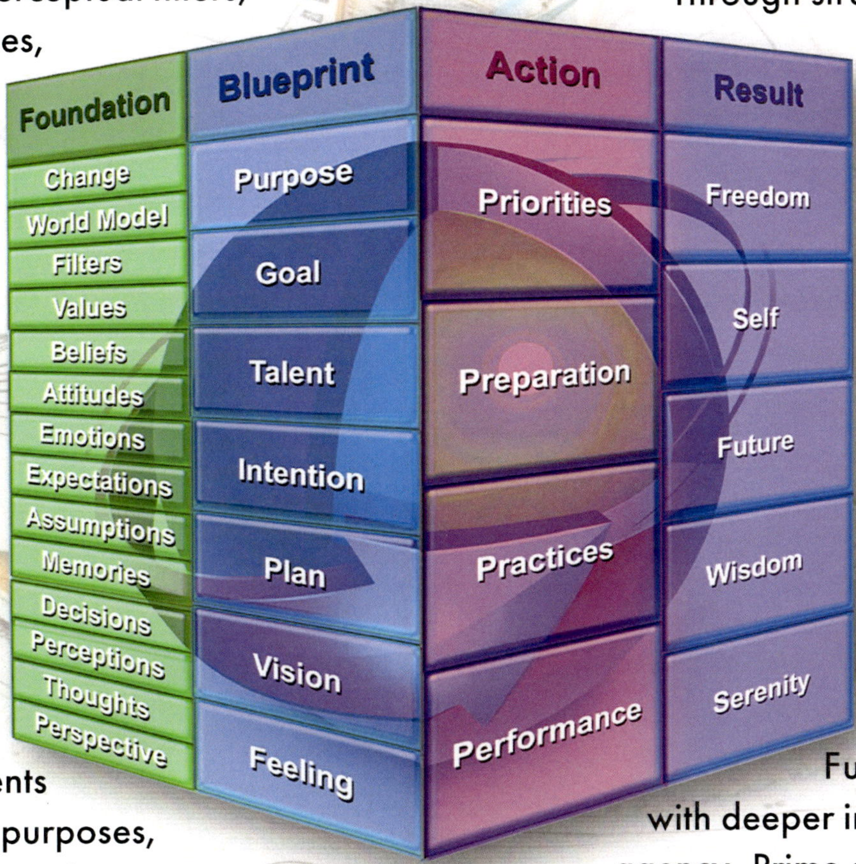

Part 2 – Blueprint: Visionary Plan: Students consciously design the purposes, goals, talents, intentions, plans, vision and emotions that will fuel their journey to Sapience. This phase emphasizes clarity and direction; ensuring that aspirations are grounded in authenticity and meaningful action.

Part 3 – Action: Empowered Journey: Primes learn to set priorities, prepare effectively, implement supportive practices and monitor their progress. Through structured strategies and adaptive learning, they cultivate discipline, adaptability and persistence; ensuring that transformation is sustainable and effective.

Part 4 – Result: Fulfilled Self: Equipped with deeper insight and empowered agency, Prime graduates are ready to experience life with greater clarity, purpose and fulfillment. The culmination of this journey is a life of freedom, self-awareness, possibility, wisdom and serenity.

Level 300 – Sapient I

Sustained Enhancement

Level 300 will deepen an individual's reservoir of conscious awareness and amplify desirable cognitive and emotional traits to elevate the person's quality of life and ability to contribute to society.

Sapient RNA therapies will be delivered incrementally to gradually lift individuals into manageable states of greater awareness and mental acuity. The therapies will be effective for 6 months.

Professional educators will develop a complete Sapient curriculum to convey key life management skills.

- *Success strategies:* goal setting, intentionality, and visualization
- *Personal productivity techniques:* priority setting and time management
- *Cognitive flexibility skills:* focus, concentration, imagination, creativity, memory, and mindfulness techniques
- *Relationship competencies;* communication, interpersonal harmony and emotional balance
- *Personal growth:* Character development, cultivating wisdom and Source connection
- *Wellbeing:* diet, nutrition, exercise, sleep, relaxation and biorhythms.

Level 400 – Sapient II

Permanent Enhancement

Level 400 will be released once gene therapies have become mainstream. At that time, eligible Sapients may apply to receive a DNA treatment for permanent neuronal genotype upgrade.

Since Sapients will be accustomed to cognitive enhancement from RNA therapy, the DNA treatment is not expected to produce noticeable changes. The DNA edit will not affect the germ line, so progeny will not retain the enhancements.

Cognitive enhancement will be performed gradually in a series of stages to ensure uniform outcomes, with results verified by neuroimaging systems and psychological tests. Campus staff will conduct regular follow-ups with each student to monitor their progress and answer questions.

Cognitive upgrades are expected to be transformative for many people. Sapients may experience positive shifts in awareness, personality, outlook, values, relationships, and professional interests.

Sapients may return to campus for continuing education programs to refine their consciousness management skills and join other graduates for networking and support. Participants will share their life experiences with others in classroom settings and receive expert guidance on optimizing their performance.

Local community alumnae groups will help Sapients navigate changes, integrate growth, exchange insights, and remain connected with their peers.

Restoring Natural Abilities

Sapience Value

The Sapient transformation can unlock a constellation of 12 mental and emotional benefits which accompany greater conscious awareness.

Many of these desirable traits will arise from Sapients having greater cognitive capital at their disposal. Others are a result of the upgrade's unique mechanisms of action in neuroanatomy.

These benefits will cascade through multiple neural systems, reinforce one another, and compound over time.

HTR2A

Master Control Switch

The 12 Sapience benefits can be generated by changing the expression of a single gene.

Gene HTR2A is a master control switch that regulates how the brain's cortical and limbic circuits process cognitive information.

This gene's expression can be modified to adjust the electrical properties of neurons to optimize mental performance and emotional balance.

These changes can yield 5 clusters of interconnected and mutually reinforcing virtuous cognitive traits.

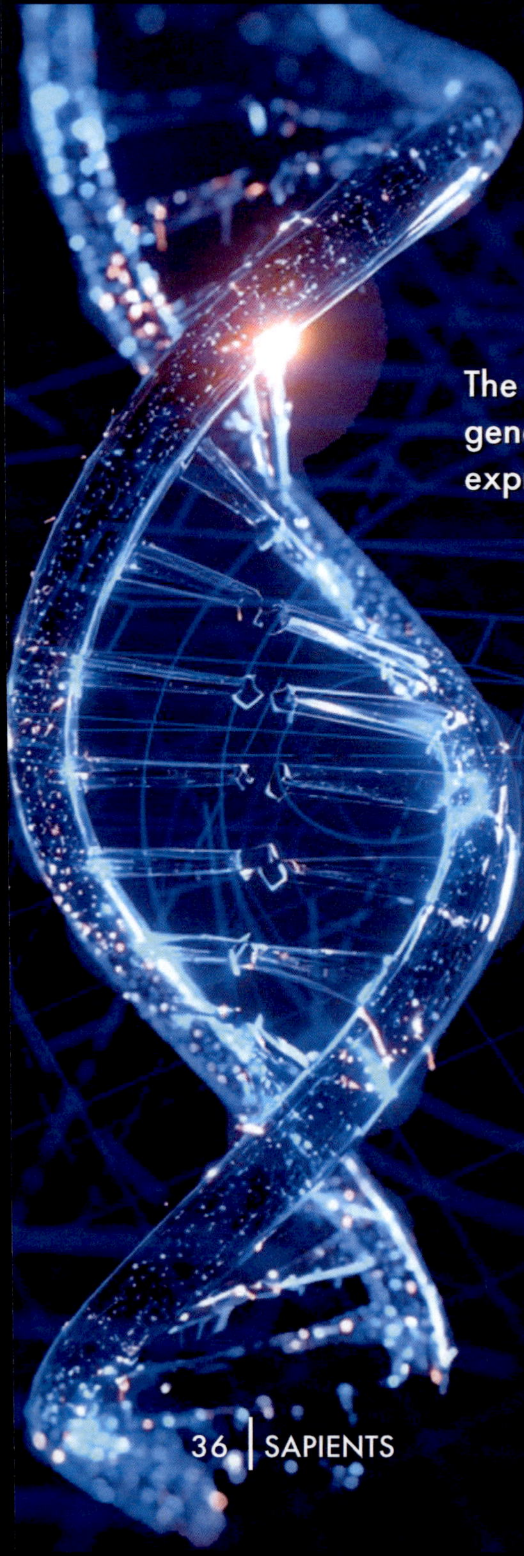

Sapience Cognitive Traits

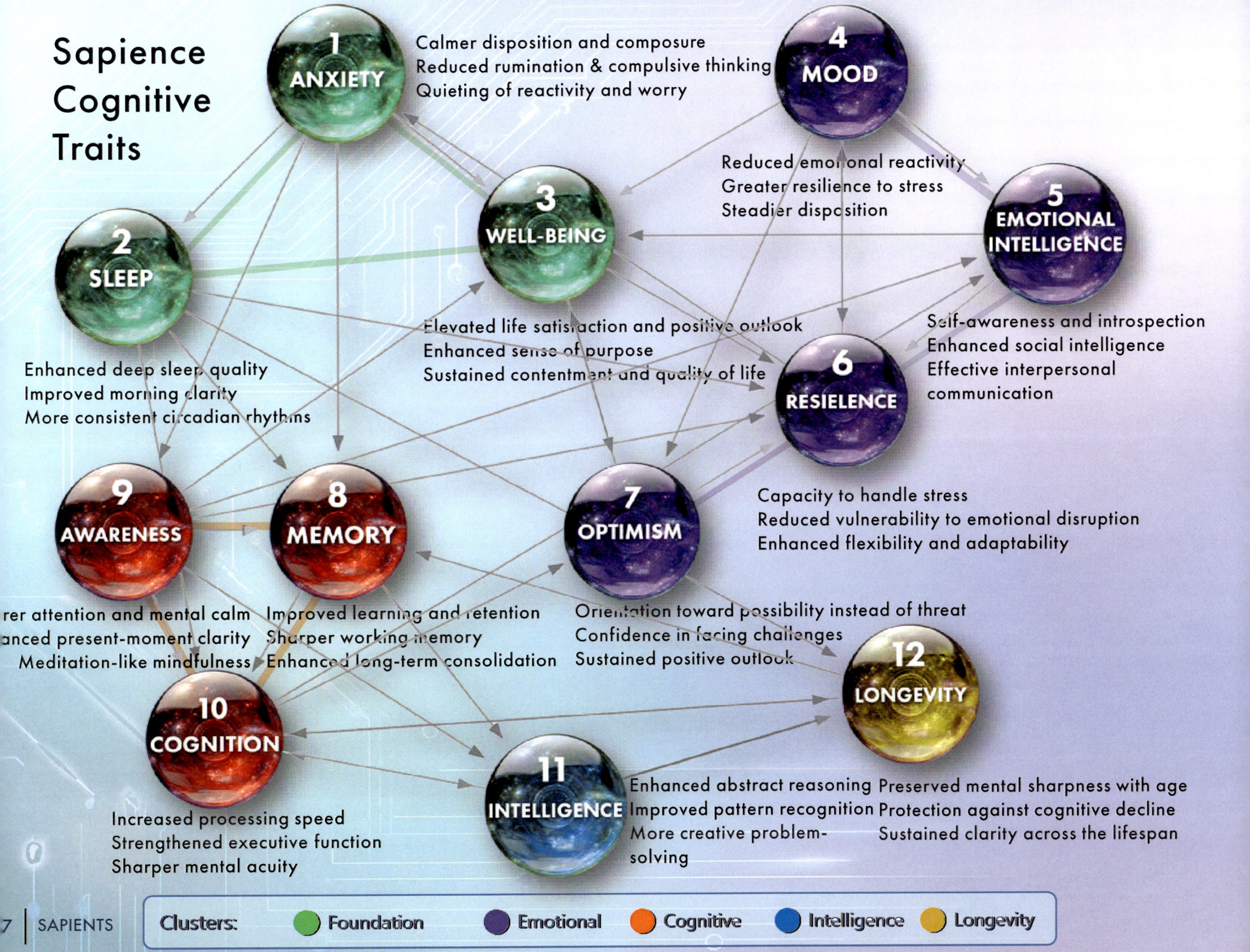

1	Reduced Anxiety	
	Affected by: Resilience	Affects: Sleep, Awareness, Memory, Well-Being

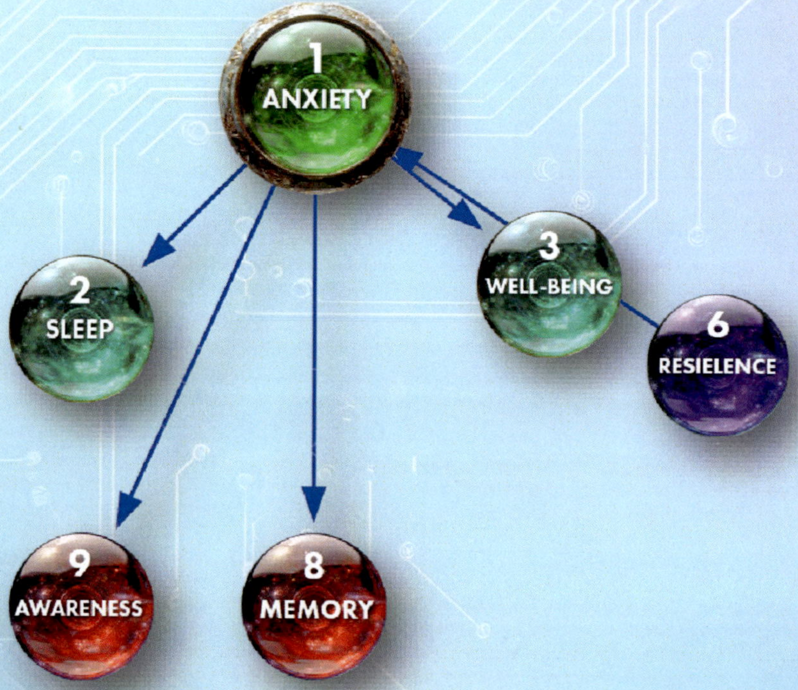

Preclinical studies downregulating HTR2A expression show a reduction in anxiety-related behaviors without sedative effects on motor function or alertness.

The improvement reflects a genuine quieting of the neural circuits that generate excessive worry, rumination, and threat vigilance.

2	Improved Sleep	
	Affected by: Anxiety	Strengthens: Memory, Cognition, Emotional Resilience, Longevity

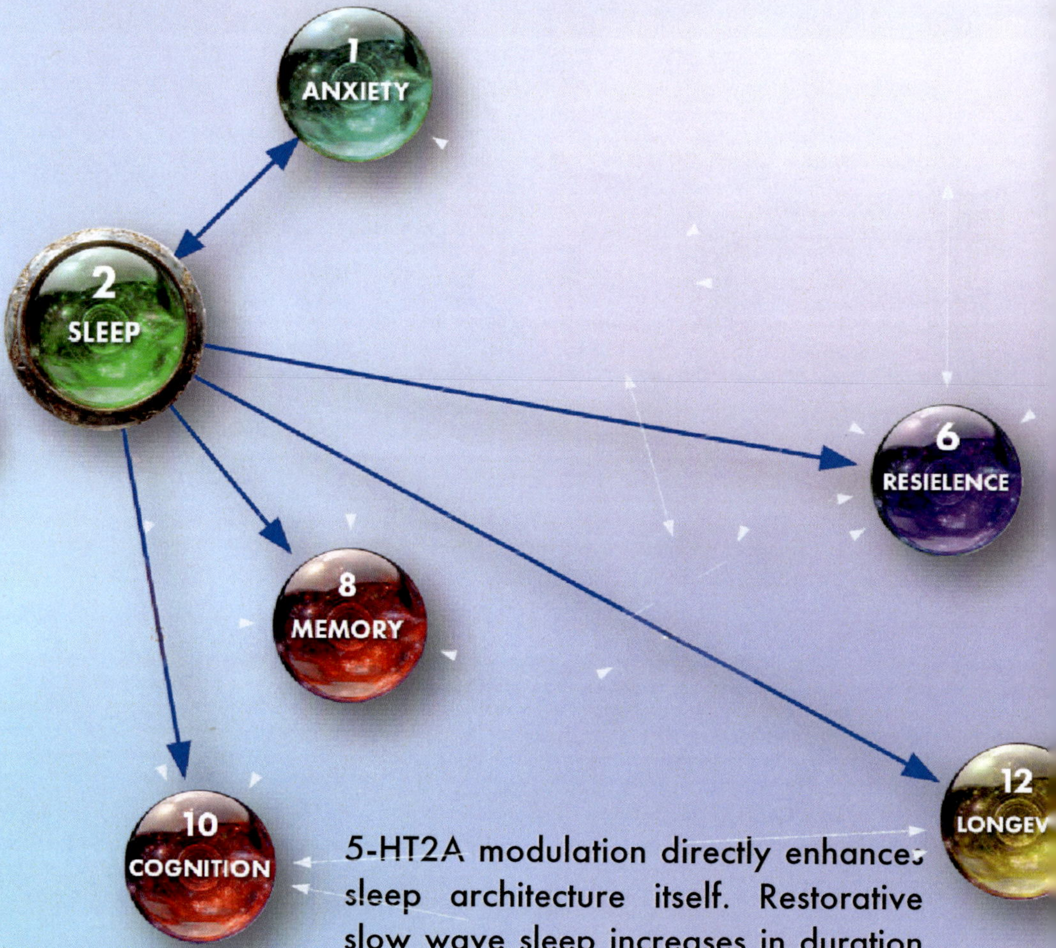

5-HT2A modulation directly enhances sleep architecture itself. Restorative slow wave sleep increases in duration and quality, improving memory consolidation and cellular repair. Nighttime awakenings decrease and morning cognition improves.

3	Enhanced Well Being	
	Affected by: Anxiety, Mood, Emotional Intelligence, Optimism, Awareness	Improves: Optimism, Resilience

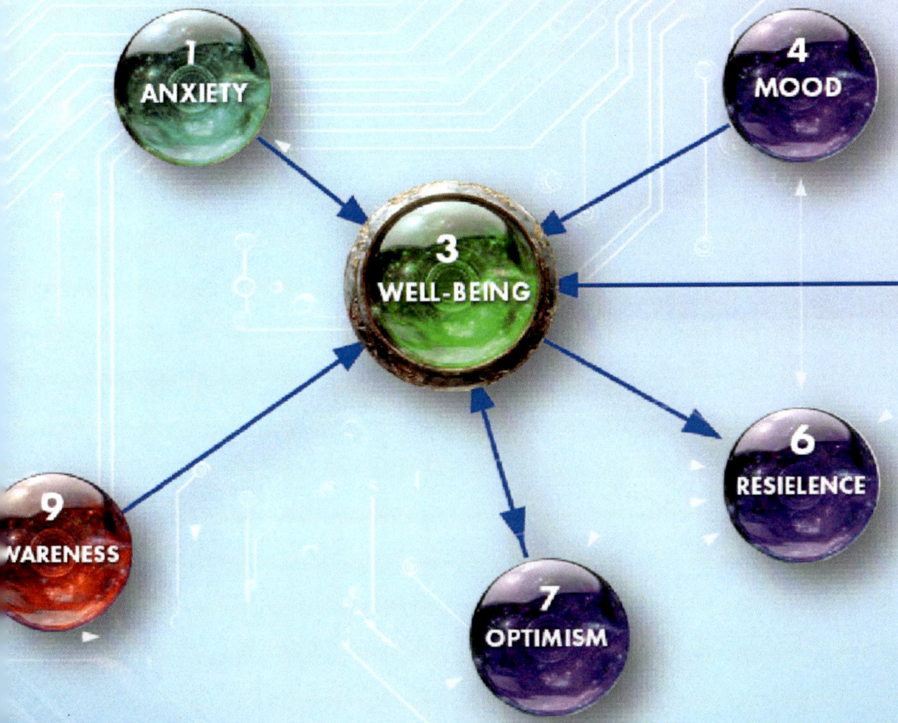

Well-being encompasses life satisfaction, positive outlook, and sense of purpose. Cognitive neuroengineering addresses the neural substrates of well-being directly, producing persistent improvements from genuine changes in brain function.

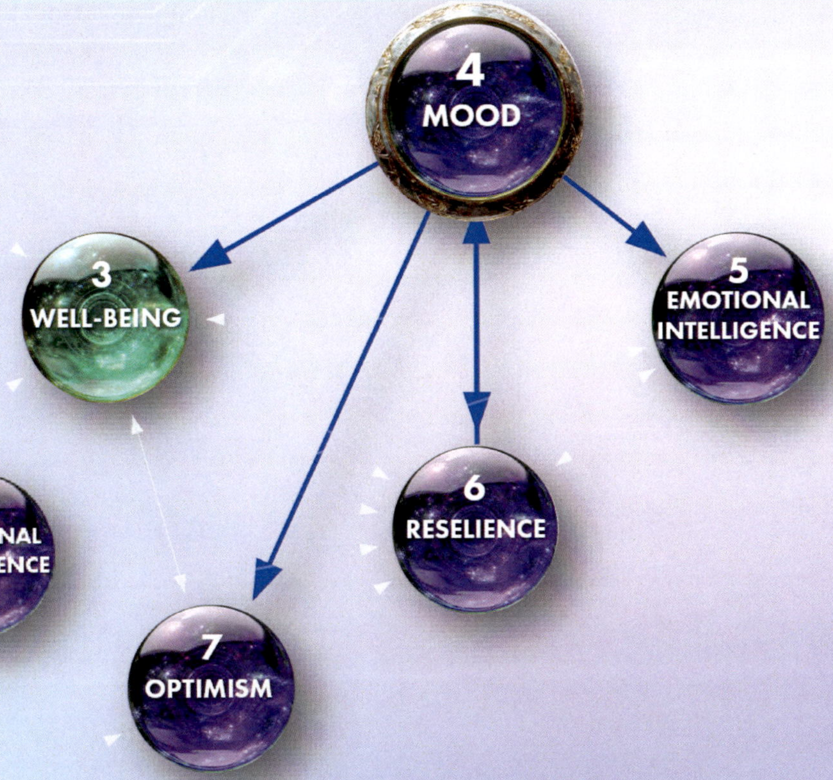

4	Stabilized Mood	
	Affected by: Resilience	Enhances: Emotional Intelligence, Resilience, Optimism, Well-being

Multiple neuroimaging studies correlate depression with the neural characteristics of elevated HTR2A expression. Conventional antidepressants address this issue indirectly and imprecisely by changing global brain chemistry, often causing intolerable side effects.

Cognitive neuroengineering can produce results directly with great precision, stabilizing mood, increasing resilience and decreasing reactivity.

5	Emotional Intelligence	
	Enhanced by: Mood, Awareness, Cognitive Function	Strengthens: Resilience, Well-being

The ability to perceive, understand, and regulate personal and interpersonal emotional states is the hallmark of emotional intelligence.

This capacity depends on the limbic system functioning within optimal parameters.

Chronic limbic hyperactivation, characteristic of anxiety and mood disorders, impairs emotional discernment.

Normalizing limbic activity restores the nuanced emotional perception that underlies social intelligence and interpersonal effectiveness.

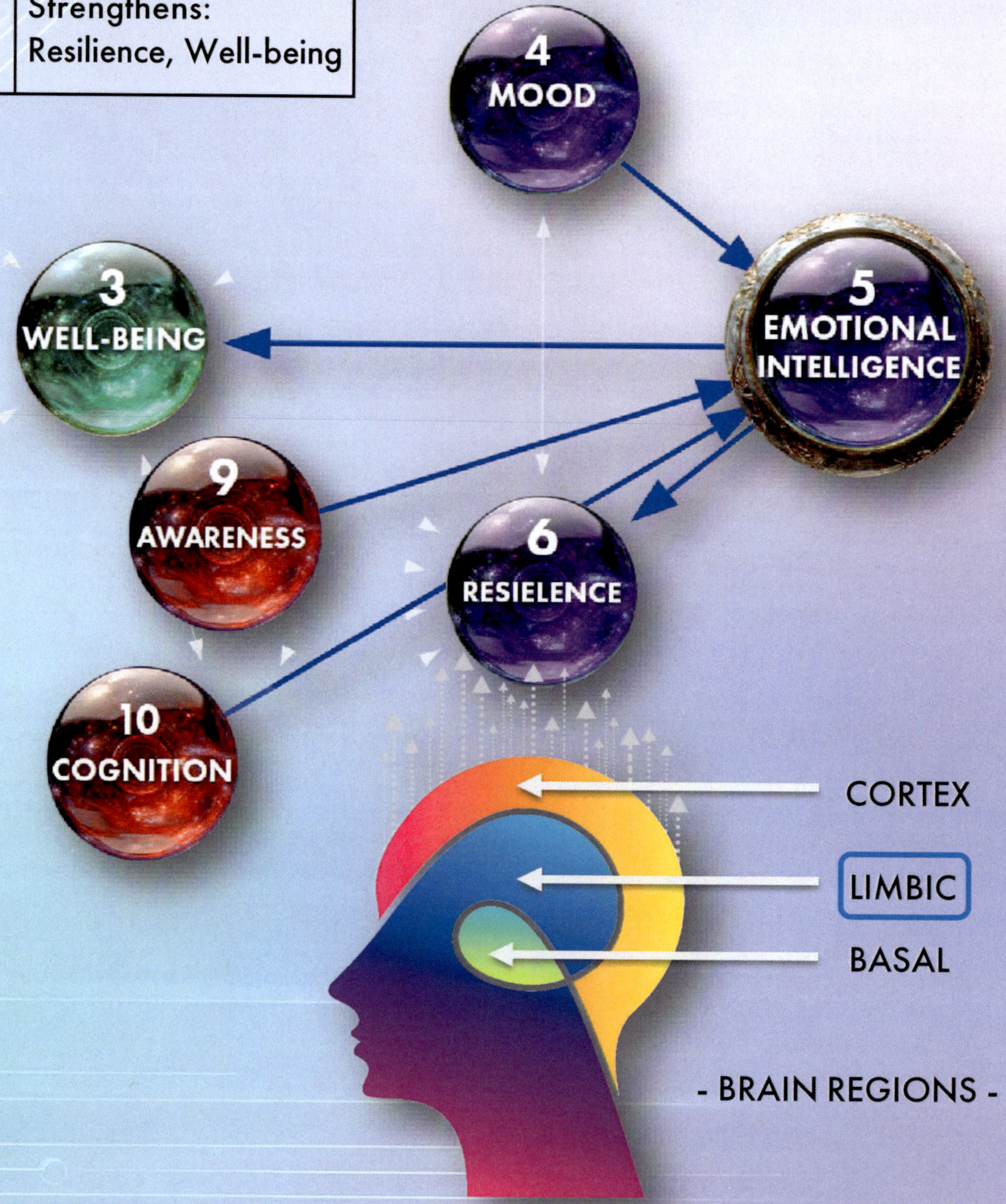

6	Fortified Resilience	
	Enhanced by: Sleep, Well-being, Mood, Emotional IQ, Optimism, Awareness	Effects: Reduces Anxiety, Stabilizes Mood

7	Increased Optimism	
	Enhanced by: Well-being, Mood, Cognition	Affects: Well-being, Resilience

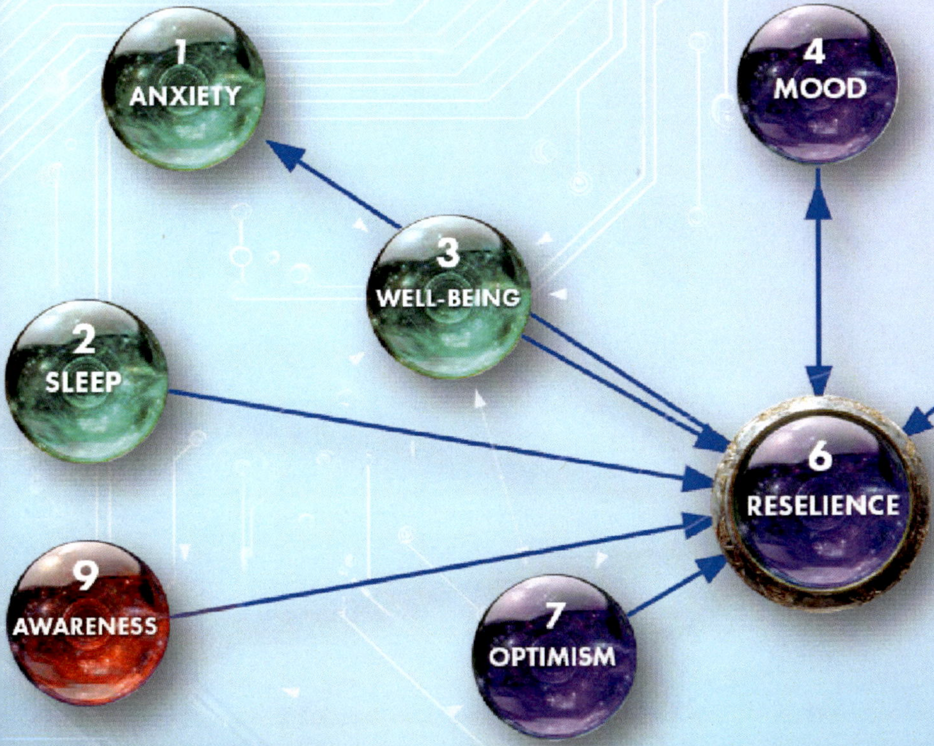

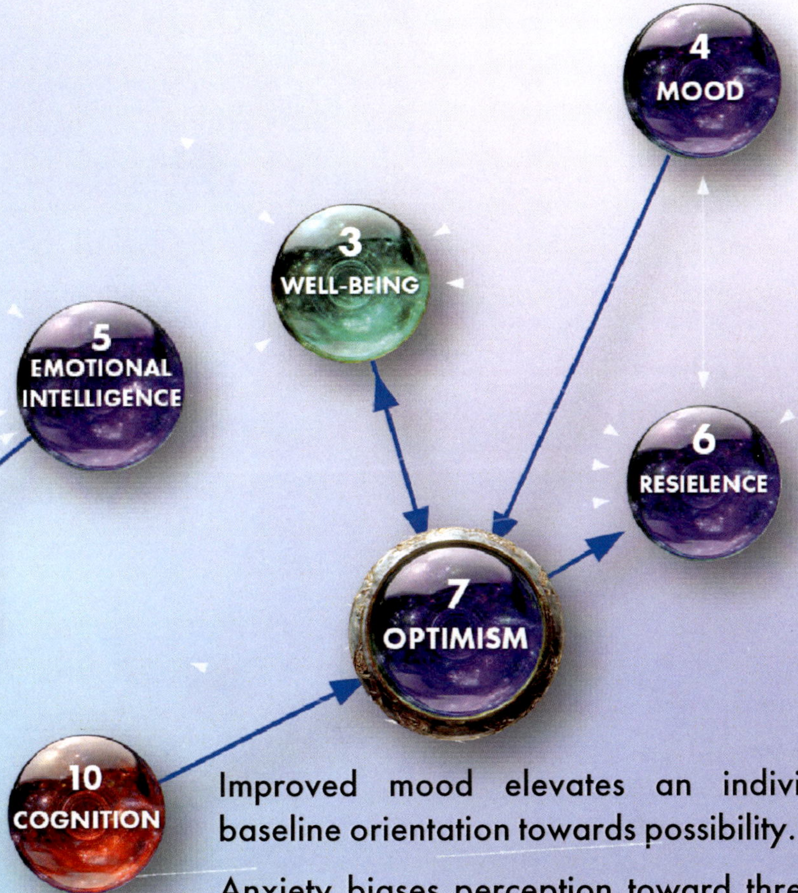

Resilience reflects the capacity to encounter stress while maintaining equilibrium.

Cognitive neuroengineering can strengthen resilience through 3 mechanisms which promote flexibility and adaptability:

a) reduced baseline anxiety lowers vulnerability.

b) normalized stress hormones prevent cortisol damage;

c) improved sleep enhances recovery from daily challenges.

Improved mood elevates an individual's baseline orientation towards possibility.

Anxiety biases perception toward threat; its reduction permits a more balanced appraisal of circumstances.

Enhanced cognitive function supports the problem-solving that transforms challenges into opportunities.

Optimism emerges naturally when the neural obstacles to its expression are removed.

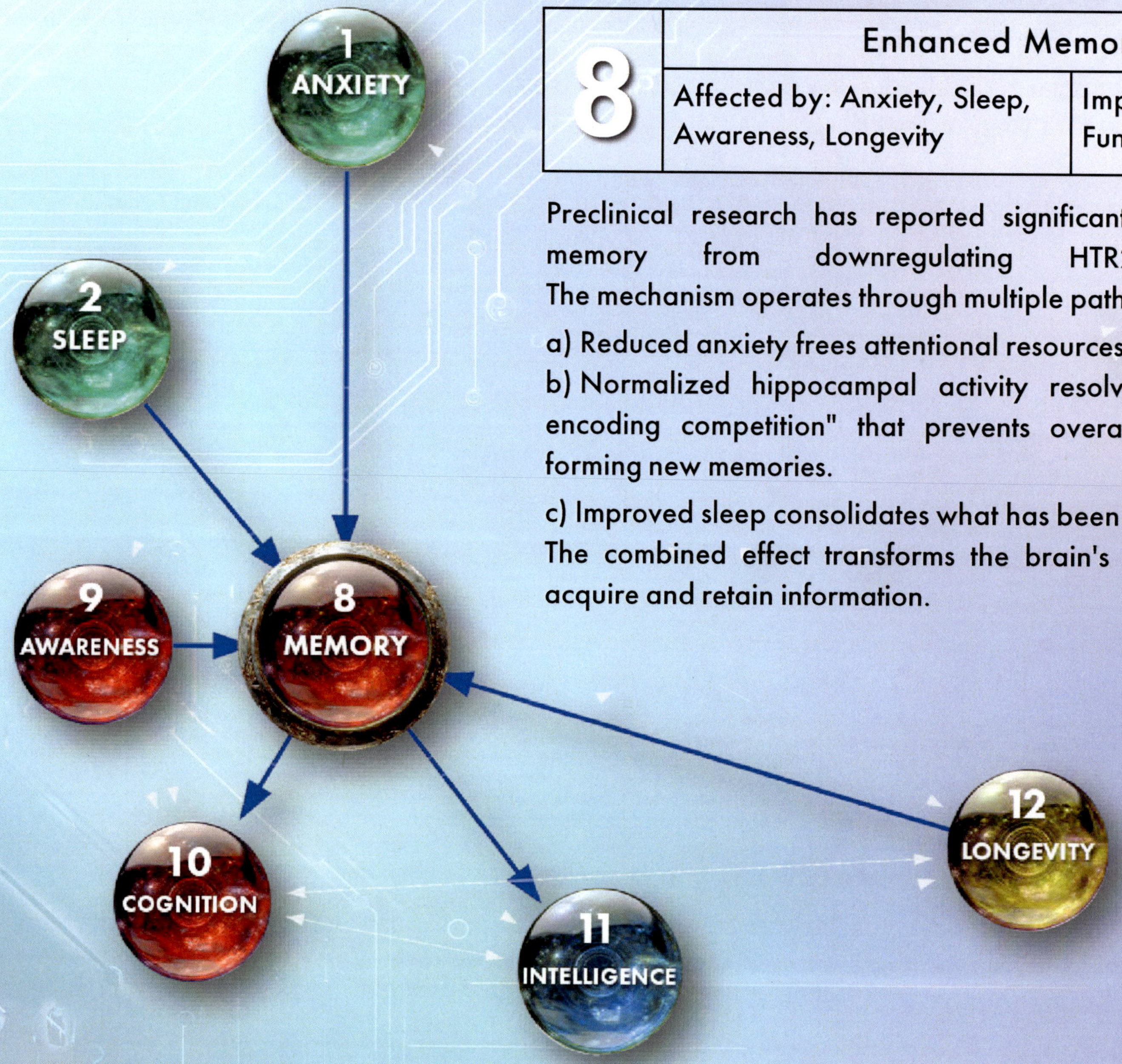

8	Enhanced Memory	
	Affected by: Anxiety, Sleep, Awareness, Longevity	Improves: Cognitive Function, Intelligence

Preclinical research has reported significant improvements in memory from downregulating HTR2A expression. The mechanism operates through multiple pathways:

a) Reduced anxiety frees attentional resources for encoding.

b) Normalized hippocampal activity resolves the "retrieval-encoding competition" that prevents overactive brains from forming new memories.

c) Improved sleep consolidates what has been learned.

The combined effect transforms the brain's capacity to acquire and retain information.

9	Heightened Awareness	
	Affected by: Anxiety	Affects: Memory, Cognition, Intelligence, Emotional IQ, Resilience, Well-being.

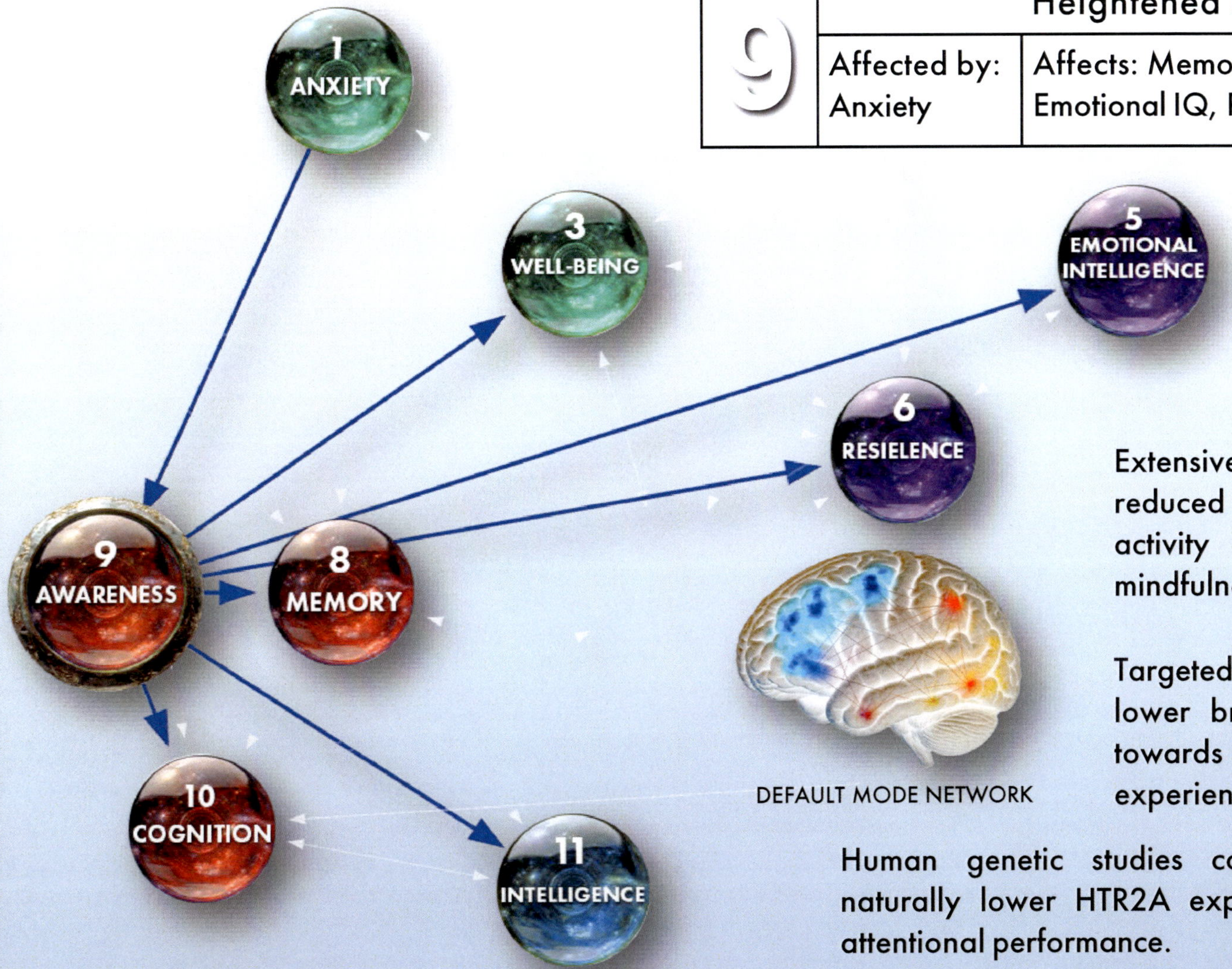

Extensive meditation research shows reduced default mode network activity correlates with enhanced mindfulness and metacognitive clarity.

DEFAULT MODE NETWORK

Targeted HRT2A downregulation can lower brain activity in this network towards the patterns characteristic of experienced meditators.

Human genetic studies confirm that individuals with naturally lower HTR2A expression demonstrate superior attentional performance.

NEURON RECEPTORS

Gene HTR2A controls the expression of a neuron receptor called 5-HT2A which modulates signal-to-noise ratio in neural circuits. When this receptor's activity is elevated, neural noise increases and attention diminishes. Reducing this activity by downregulating HTR2A can produce cleaner signals, better filtering of distractions, and improved focus.

10 Cognitive Function

Affected by: Sleep, Memory, Awareness, Intelligence, Longevity	Affects: Intelligence, Emotional IQ, Longevity, Optimism

HTR2A modulation can improve overall cognitive efficiency on a global level, increasing processing speed, strengthening executive function and expanding verbal fluency.
Large-scale human studies of medications which block or reduce 5-HT2A receptor expression report improvements across multiple cognitive domains, with effects sustained over years of treatment.

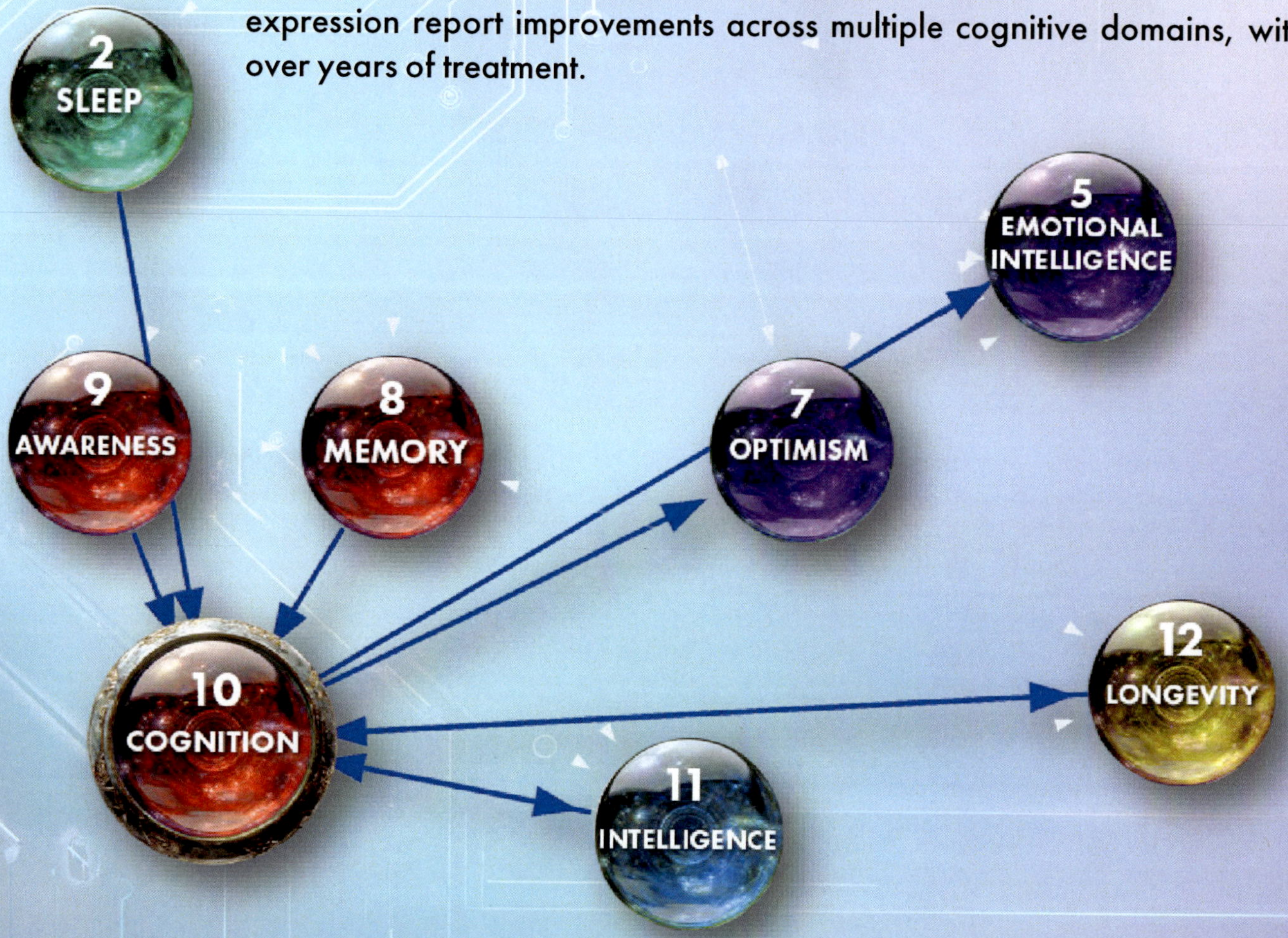

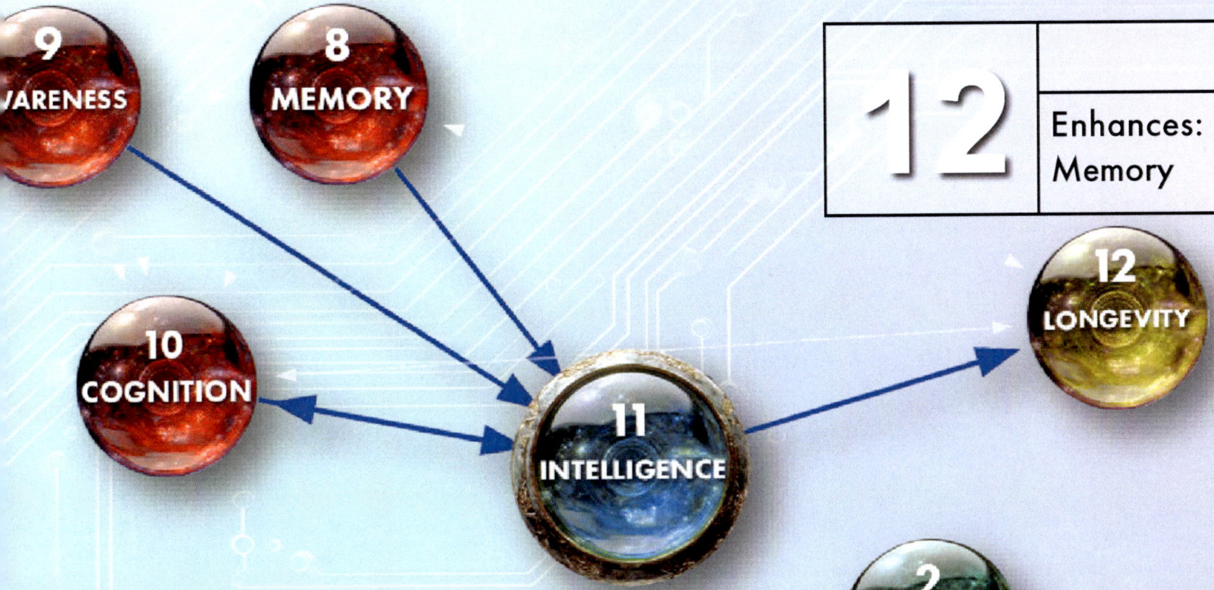

12	Cognitive Longevity	
	Enhances: Cognition, Memory	Supported by: Sleep, Cognition, Intelligence

Reduced neural excitation is a key factor in cognitive longevity. Neural hyperactivity accelerates cognitive decline in humans, and interventions that attenuate it extend lifespan in model organisms.

Targeted HTR2A downregulation aligns with these fundamental mechanisms, positioning cognitive longevity as a predictable outcome of reduced neural excitation.

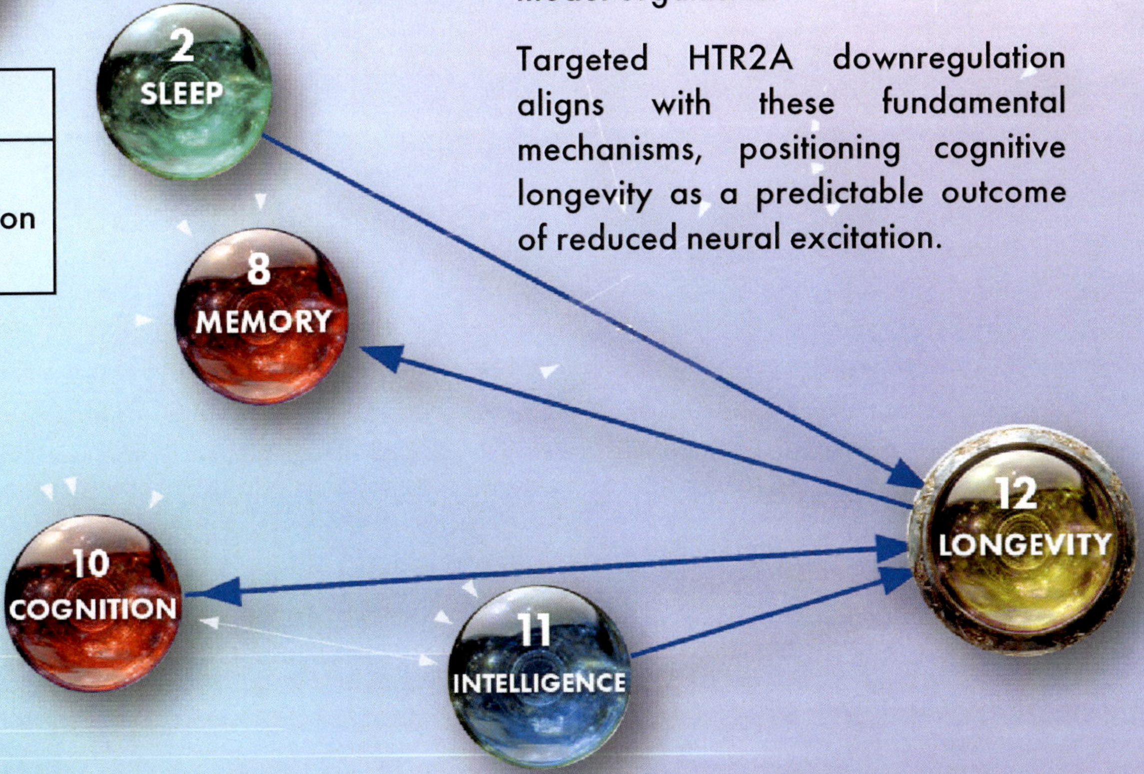

11	Higher Intelligence	
	Enhanced by: Memory, Awareness, Cognition	Supports: Cognitive Function and Longevity

Intelligence correlates with the neural capacity to efficiently accomplish cognitive work with minimal metabolic expenditure.

Higher intelligence is associated with lower neural dendritic density and reduced glucose consumption during cognitive tasks.

Selectively reducing HTR2A expression can optimize neural architecture toward this efficient configuration, enhancing the brain's capacity for abstract reasoning, pattern recognition, and problem-solving.

Why Sapients? Why not just a menu of upgrades?

Identity-level change is the most potent lever in personal transformation. When a person shifts who they *are,* not just what they *do*, the effects cascade through every domain of their life.

Modifying neuronal genetics creates a golden opportunity to affect this level of change.

Genomic adjustments are not metaphors or affirmations—they are hard wired changes in biology. Consequently, they confer an authentic license to reframe identity.

This approach has much greater impact than offering a menu of cognitive upgrades to be consumed like pills. In that model, the catalytic power resides in an external product, not in the individual. The person becomes a consumer of improvement rather than its author.

Anchoring the upgrade in a new self-image internalizes its advantages, allowing. Sapients to claim ownership of their expanded capacities. It also creates downstream benefits, including the amplifying effects of belief and expectation.

4.1: Sapient Community

In its ideal form, a Sapient community will be bound by shared values, including authenticity, integrity, and ethical leadership. These common principles will foster a culture of mutual respect and collaboration.

The journey to Sapience will foster a profound shift in identity. Sapients will view themselves as pioneers of a new human epoch, driven by purpose and responsibility. This elevated sense of role will encourage them to act as stewards of humanity's collective future and leaders in their respective fields.

By aligning individual goals with collective action, the Sapient community has the potential to reshape industries, influence global governance, and redefine what it means to thrive in the modern world.

4.2: Social Impact

The Cognitive Age

4.2.1: New Skills for the AI Era

As society migrates to an AI economy, the foundations of human livelihood are shifting.

Human traits rewarded in the Industrial and Information Ages are giving way to new skills that will define the Cognitive Age.

Pre-AI Economy	AI Economy
Industrial + Information Age	*Cognitive Age*
Speed, compliance, endurance	Imagination, insight, presence
Memorization, repetition	Empathy, curiosity, intuition
Linear problem-solving	Cross-domain synthesis, originality
Technical mastery	Self-awareness, adaptability

The AI era requires more imagination, creativity, and human connection. Society will increasingly prioritize these inner qualities over external productivity.

4.2.2: The Changing Nature of Work

Now that developed nations have entered the AI era, companies are shedding millions of jobs.

In past transitions from agriculture to industry, or from industry to services, new domains of purpose, value, and economic activity arose, even if they were hard to predict in advance.

Similarly, the AI economy is fostering entirely new human enterprises, such as:

- Purpose-based microenterprises and creator guilds
- Consciousness and well-being industries
- Regenerative environmental work
- Experience economy (beyond entertainment)
- Exploration (space, deep sea, consciousness)
- Scientific and artistic frontier work

In short, livelihoods will continue to evolve, even if "jobs" as we know them are shrinking.

The Cognitive Age

4.2.3: Strategic Cognitive Skills

Trait	How It Adds Value in the New Economy
Imagination	Designing new realities, businesses, lifestyles, or art forms that AI cannot originate.
Creativity	Synthesizing across domains to generate novel ideas, solutions, and visions.
Metacognition	Monitoring and improving one's own thinking—essential for innovation and self-directed learning.
Emotional Intelligence	Navigating human dynamics in caregiving, leadership, teaching, and community building.
Interpersonal Resonance	Deep listening, mentoring, coaching, spiritual guidance—areas AI cannot replicate.
Intuition	Making meaningful decisions under uncertainty, where data is incomplete.
Presence & Awareness	Vital for roles in well-being, mindfulness, coaching, and the inner development economy.

Cognitive traits reflecting uniquely human abilities that artificial systems cannot replicate will define success in a world where intelligence is ambient and ubiquitous.

4.2.4: Socio-Economic Benefits

Area	Enhancement Benefit	Strategic Outcome
Education & Retraining	Faster skill acquisition	Easier transition to new livelihood domains
Microenterprise & Innovation	Original ideas and clarity	Rise of high-impact, self-created livelihoods
Community Leadership	Emotional balance + understanding	Local governance, mediation, and trust-based economies
Creative Professions	Fluidity + imagination	Dominance in art, design, media, experience curation
Self-Healing & Well-Being Roles	Self-awareness + presence	Emergence of a scalable inner development sector

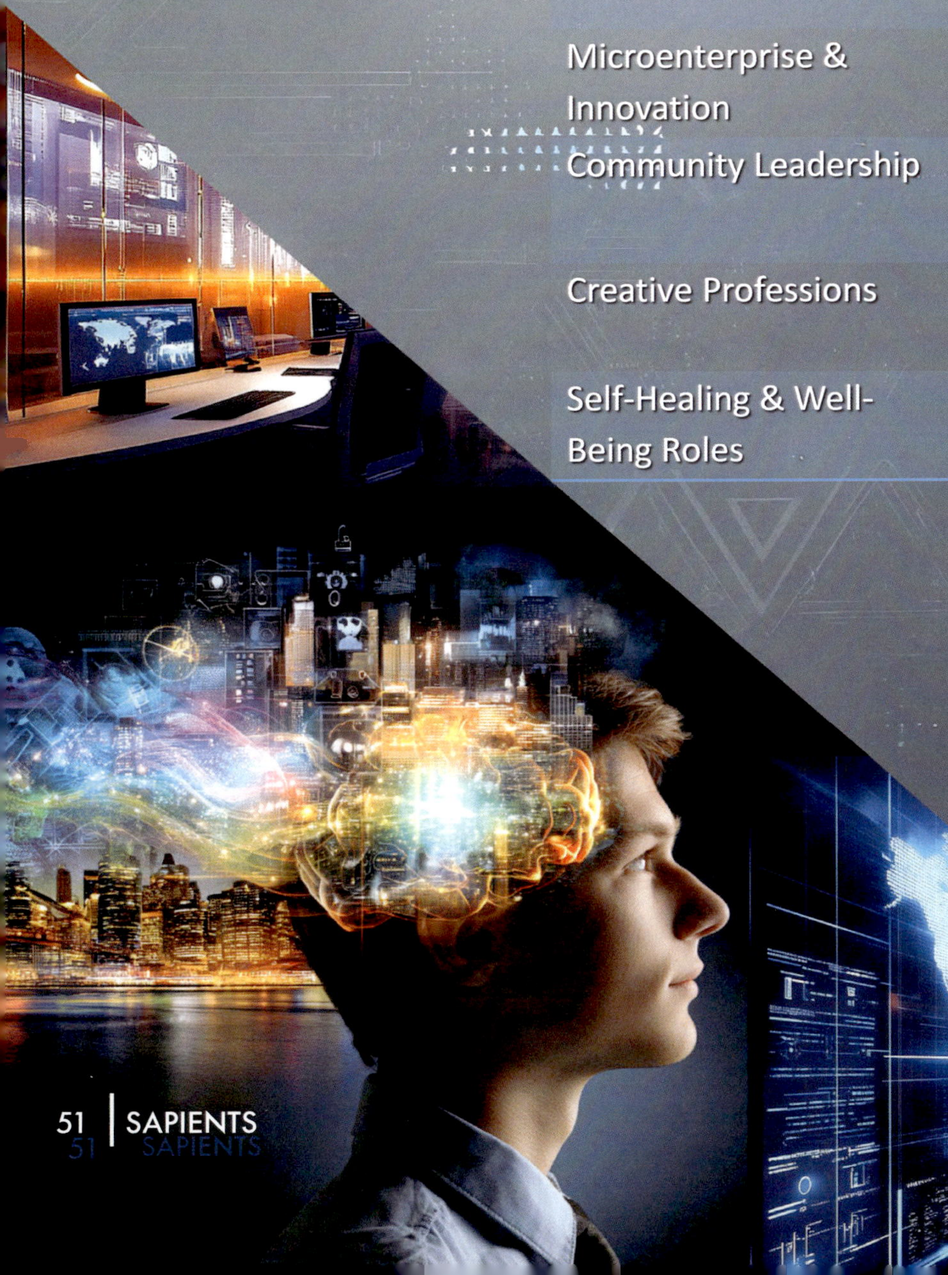

Economic benefits accrue from desirable cognitive competencies, and cognitive upgrades will become essential for social and economic transition.

4.2.5: Human Capital Transformation

Genetic and behavioral cognitive enhancement technologies can unlock this new class of traits by accelerating learning, deepening interpersonal connection, and supporting self-mastery. For example: →

Once a privilege, literacy is now a human right. In the future, cognitive literacy and enhancement may be treated the same way: a public good needed to unlock society's full potential.

- Increased cognitive flexibility for rapid upskilling
- Reduced anxiety for clearer decision-making
- Improved focus for creative and strategic work
- Heightened attunement for collaborative or healing roles
- Access to flow states for fulfillment and productivity

Investing in technologies and platforms that promote mental acuity, emotional intelligence and self-awareness can build a human human cognitive substrate that will support thriving in the AI economy.

4.3: Global Impact

Sapients will be uniquely equipped to tackle humanity's most pressing challenges. Their enhanced cognitive and emotional abilities will support them in leading sustainable development, resource management, and equitable economic growth. By deploying cognitive capital into these areas, Sapients can serve as catalysts for systemic change.

Cognitive enhancement can catalyze conscious values and new behaviors that are more closely aligned with holistic ecology models and sustainable development goals.

At high levels, this change can shift capital flows to collateralize deep impact initiatives supporting these objectives.

The broader implications of Sapience extend to global challenges.

To address climate change and other existential threats, humanity requires a level of consciousness current methods cannot achieve at scale.

Sapients represent a scalable solution — a community capable of thinking and acting with unprecedented clarity, creativity, and moral insight.

4.4: Legacy

Sapience holds the promise of humanity's next great chapter. By unlocking latent cognitive capacities, Sapients can pave the way for achieving milestones that were once thought impossible.

Colonizing the solar system, resolving global conflicts, and achieving ecological balance are no longer unattainable dreams but tangible goals within reach.

4.5: Catalyzing Cognitive Change

In a world where the boundaries of human cognition are being redrawn, we stand at the threshold of a cardinal transformation:

Current genetic technology coupled with the neuro-modulation techniques described in this book can recalibrate neural networks to unlock dormant cognitive potential.

These breakthroughs demand a translation into practice. The time has come to mobilize the expertise and build the infrastructure required to elevate human cognition. A strategic plan for progress follows on the next page.

THE SAPIENT INITIATIVE

1. *Form a Board of Trustees:* Assemble a visionary board comprising leaders with multidisciplinary skills to guide the nascent Cognitive College through the intersections of technology, education and policy.

2. *Develop Policies:* Draft robust policies to govern research ethics, data security, college administration, faculty recruitment, admissions and clinical applications.

3. *Hire Staff:* Recruit a dedicated team of scientists, clinicians, educators, and administrators who share the vision of a cognitive renaissance.

4. *Advance Genetic Technology:* Invest in leading-edge research to transition cognitive gene therapies from the laboratory and clinic into the public domain. Continuously refine genetic methods for optimizing cognitive function.

5. *Build an Online Community:* Develop a state-of-the-art web platform to serve as an interactive forum for global collaboration, continuous dialogue, data sharing and rapid dissemination of breakthroughs.

THE SAPIENT INITIATIVE

6. Initiate Conferences: Launch a series of high-profile conferences that bring together influential thought leaders, researchers and policymakers. These gatherings will disseminate knowledge and foster the relationships necessary for rapid innovation.

7. Develop a Curriculum: Begin crafting an interdisciplinary curriculum, starting with the education program and therapeutic protocols for Level 100: Whole.

8. Build a Blockchain: Establish a secure cognitive enhancement blockchain registry to provide auditability and quality control by tracking manufacturing, clinical delivery and user outcomes.

9. Issue a Token: Democratize access to the technology with a token, allowing early adopters and micro-investors to benefit as the token appreciates. Create a donor-supported scholarship fund which tokenizes member participation in conferences and events.

10. Buy Land and Break Ground: Secure a physical campus to serve as the epicenter of cognitive innovation where research, education and practical application converge to create transformative experiences.

59 SAPIENTS

4.6: Leading Transformation

Influential individuals, policymakers, and thought leaders are the prime movers in this grand vision. Their engagement is crucial in three ways:

Investing in the Vision:

Allocating intellectual and financial resources to research and development programs to bring these ideas from the lab to everyday life.

Championing Change:

Advocating for policies that support cognitive enhancement research and the ethical application of genetic neuro-modulation technologies.

Leading by Example:

Participating in inaugural conferences, contributing to curriculum development, and building the leadership network that will redefine the future of human cognition.

A Defining Moment

he journey from engineering theory to ractical application is often hallenging, yet the rewards here are mmense. Embracing these initiatives an usher in a new era where cognitive nhancement is not a luxury, but a ornerstone of human evolution.

The scientific evidence and visionary insights detailed in this foundational text provide both inspiration and direction. The blueprint is ready—now is the time to build, innovate, and transform.

Working together, we can break new ground in the science of consciousness and lay the foundation for a future where human potential is limitless.

COGNITIVE COLLEGE

Course Catalog & Curriculum Guide

LEVEL 100 · WHOLE *Balance & Resilience*

Modules	Track	Description
WHL 101–105	Mental Equilibrium Core	Psychological assessment, stress modulation, emotional regulation, anti-fragility, and therapeutic integration to establish baseline stability.
WHL 110–114	Physiological Optimization	Sleep architecture, nutritional neuroscience, movement, biorhythm synchronization, and breathwork to optimize the body for cognition.
WHL 120–124	Consciousness Foundations	Introduction to awareness science, meditation fundamentals, attention training, daily mindfulness, and cultivating a quiet mind.

LEVEL 200 · PRIME *Ready for Upgrade*

Modules	Track	Description
PRM 201–206	Inner Alignment	Examining perceptual filters, clarifying values, restructuring beliefs, processing emotional patterns, integrating memory and identity, and cultivating openness to transformation.
PRM 210–215	Visionary Blueprint	Purpose discovery, goal architecture, visualization, mental rehearsal, intention setting, life design, and talent mapping to form a clear vision for enhancement.
PRM 220–224	Empowered Action	Priority setting, implementation science, progress monitoring, discipline and self-regulation, and sustainable practice development to convert intention into results.
PRM 230–234	Enhancement Preparation	Science of HTR2A modulation, personal neuroimaging interpretation, protocol familiarization, transition management, and ethics of enhancement.

LEVEL 300 · SAPIENT I *Sustained Enhancement*

Modules	Track	Description
SAP 301–306	Cognitive Amplification	Focus mastery, memory optimization, creative intelligence, learning acceleration, mental clarity, and imagination development leveraging RNA-based enhancement.
SAP 310–314	Metacognitive Development	Metacognition fundamentals, deepening self-awareness, advanced mindfulness, witness consciousness training, and cognitive flexibility.
SAP 320–324	Emotional Mastery	Enhanced emotional intelligence, optimism cultivation, emotional balance, compassion and empathy development, and managing heightened sensitivity.
SAP 330–334	Interpersonal Excellence	Advanced communication, relationship dynamics, conflict resolution, leadership and influence, and collaborative team intelligence for Sapients.
SAP 340–344	Integration	Processing the enhancement experience, professional application, personal relationship navigation, life purpose actualization, and ongoing optimization protocols.

LEVEL 400 · SAPIENT II *Permanent Enhancement*

Modules	Track	Description
SAP 401–404	Permanent Enhancement Preparation	DNA vs. RNA modification science, long-range life planning, psychological preparation for irreversibility, and advanced neuroimaging assessment.
SAP 410–414	Advanced Consciousness Studies	Wisdom cultivation, existential intelligence, intuition development, source connection, and character development at the highest levels of awareness.
SAP 420–424	Societal Contribution	Applying enhanced cognition to global challenges, sustainable development leadership, ethical leadership in the Cognitive Age, community building, and policy advocacy.
SAP 430–434	Continuing Excellence	Lifelong learning design, peer mentorship certification, alumni leadership, research participation, and legacy planning for sustained impact.

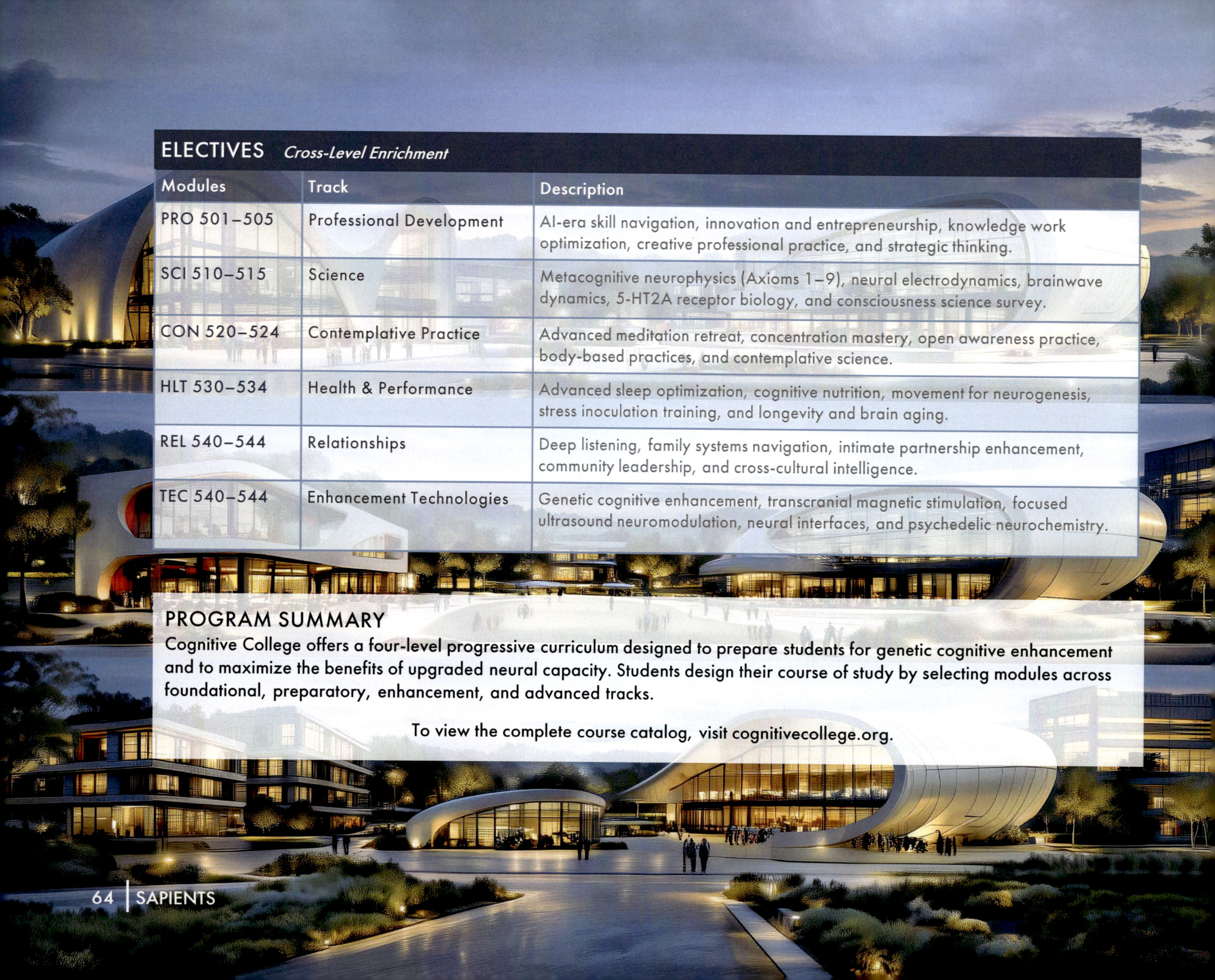

ELECTIVES *Cross-Level Enrichment*

Modules	Track	Description
PRO 501–505	Professional Development	AI-era skill navigation, innovation and entrepreneurship, knowledge work optimization, creative professional practice, and strategic thinking.
SCI 510–515	Science	Metacognitive neurophysics (Axioms 1–9), neural electrodynamics, brainwave dynamics, 5-HT2A receptor biology, and consciousness science survey.
CON 520–524	Contemplative Practice	Advanced meditation retreat, concentration mastery, open awareness practice, body-based practices, and contemplative science.
HLT 530–534	Health & Performance	Advanced sleep optimization, cognitive nutrition, movement for neurogenesis, stress inoculation training, and longevity and brain aging.
REL 540–544	Relationships	Deep listening, family systems navigation, intimate partnership enhancement, community leadership, and cross-cultural intelligence.
TEC 540–544	Enhancement Technologies	Genetic cognitive enhancement, transcranial magnetic stimulation, focused ultrasound neuromodulation, neural interfaces, and psychedelic neurochemistry.

PROGRAM SUMMARY

Cognitive College offers a four-level progressive curriculum designed to prepare students for genetic cognitive enhancement and to maximize the benefits of upgraded neural capacity. Students design their course of study by selecting modules across foundational, preparatory, enhancement, and advanced tracks.

To view the complete course catalog, visit cognitivecollege.org.

Chapter 6: First Principles

INTRODUCTION

For centuries, humans have been using meditation, psychedelics, prayer, rituals and other means to reach higher states of being. These expanded states call forth dormant awareness from our unconscious and bring it to the forefront of our experience.

Our submerged awareness constitutes an untapped reservoir of human potential which, when harnessed, can illuminate a constellation of desirable cognitive traits and abilities.

This chapter introduces a new science for achieving these aims.

AXIOM 1

Awareness is a primary universal element.

Awareness is fundamental and irreducible It is a foundational principle, not an emergent property.

A modern genetic consciousness enhancement technology must be built on the foundation of a 21st Century awareness science, not on outdated reductionist theories of the past.

Pioneering researchers have amassed incontrovertible evidence for non-local states of consciousness which clearly reveal the limits of reductionism. Dean Radin, Deepak Chopra, Edward Kelly and others have been calling for a more inclusive consciousness science for years.

The time has come to formalize a scientific discipline which treats consciousness as a first principle of nature rather than a derivative of matter.

"The day science begins to study non-physical phenomena, it will make more progress in one decade than in all the previous centuries." – Nikola Tesla

AXIOM 2

Awareness is composed of discrete packets.

In the same way that light is made of photons, awareness consists of elemental units.

Treating awareness as a granular phenomenon lays the foundation for a mathematics of consciousness.

ITONS
Conscious Awareness Packets

AXIOM 3

Awareness packets have two states.

	Conscious (ITONS)	Unconscious (OTONS)
Type	Free	Applied
Time	Present	Past
Change	Motionless	Oscillates
Behavior	Particle	Wave
Function	Being	Knowing
State	Awareness	Information

OTONS
Unconscious Awareness Packets

AXIOM 4

Awareness can change state.

Awareness packets can change state between conscious and unconscious, like a photon can shift between a particle and a wave.

AXIOM 5

Free awareness is conscious and formless.

Free awareness, comprised of conscious packets, is changeless, motionless and stationary. The Vedic Upanishads describe this formless awareness as having no mass or energy, or location in space or time.

COROLLARY 1

Conscious awareness packets have no amplitude or wavelength.

Frequency and amplitude are forms which exist in time and space. Free awareness has a non-oscillating, static nature.

This observation has implications for enhancing human cognition.

AXIOM 6

Applied awareness takes form and becomes unconscious.

Unconscious awareness takes the form of information.

Frequency is the universal carrier of information.

To hold information, unconscious oton packets must oscillate.

This gives them a wavelength, whereas conscious packets are changeless.

The Law of Conservation of Awareness

Awareness cannot be created or destroyed; only changed from one form to another.

The measurements of consciousness and unconsciousness are reciprocals of one another; as one increases, the other decreases. However, an individual's total amount of awareness remains constant, reflecting a law of conservation.

The chart uses the term iotons to represent the elemental units of consciousness, existing in either conscious (itons) or unconscious (otons) states. Iotons can change from one state to another, but their total number stays the same.

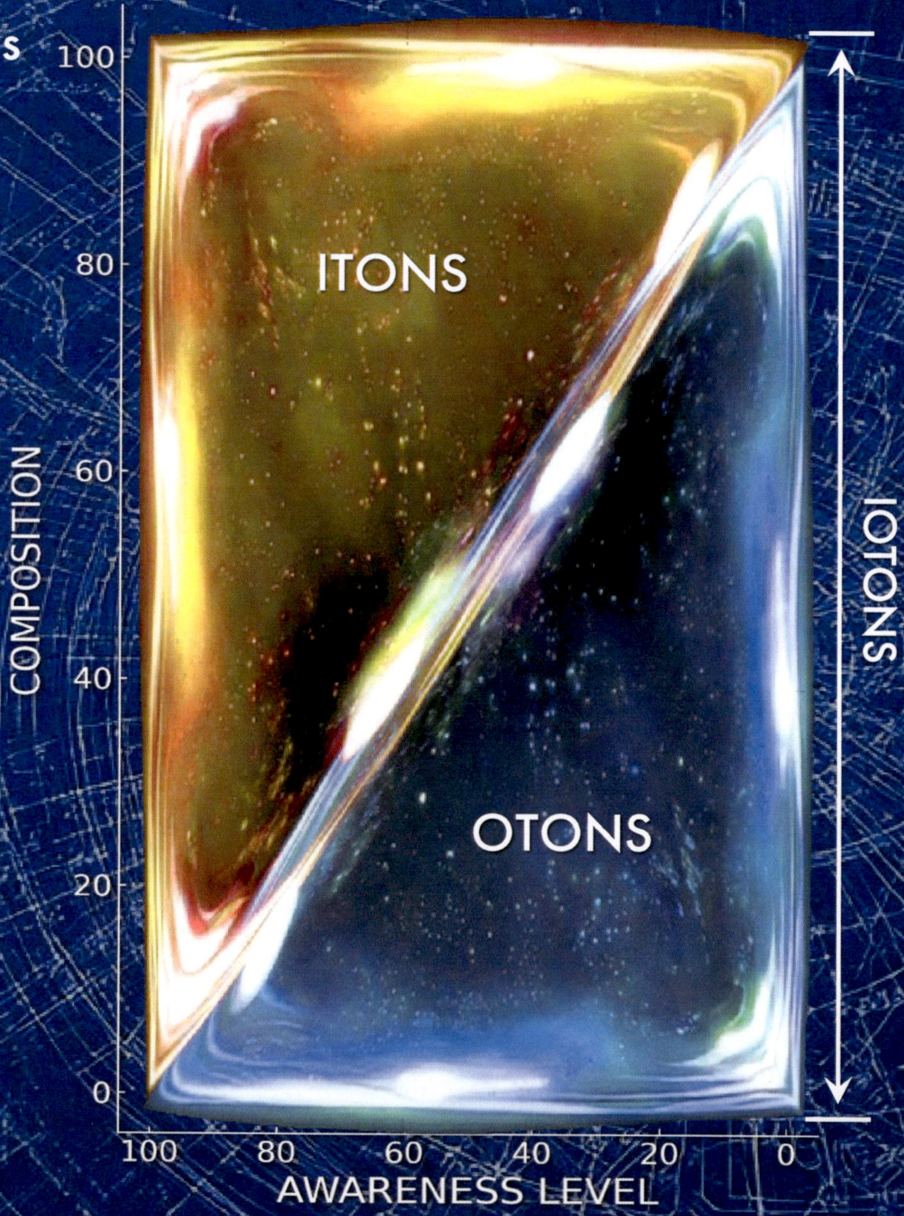

Conclusions

1. Premise

Expanded states of awareness and cognitive ability are logically associated with a higher population of conscious awareness packets.

2. Deduction

Since itons do not oscillate and have no wavelength, larger numbers of itons should be correlated with lower brainwave activity.

3. Theory

Greater awareness and cognitive capacity should therefore coincide with calmer brainwave states.

4. Evidence

A large body of scientific research indicates that shifting excess brainwave activity from high-frequency states to less active, more coherent states can unlock latent cognitive potential.

Elements of Human Cognition

Brainwaves

The Interface Zone Between Consciousness and Neurology

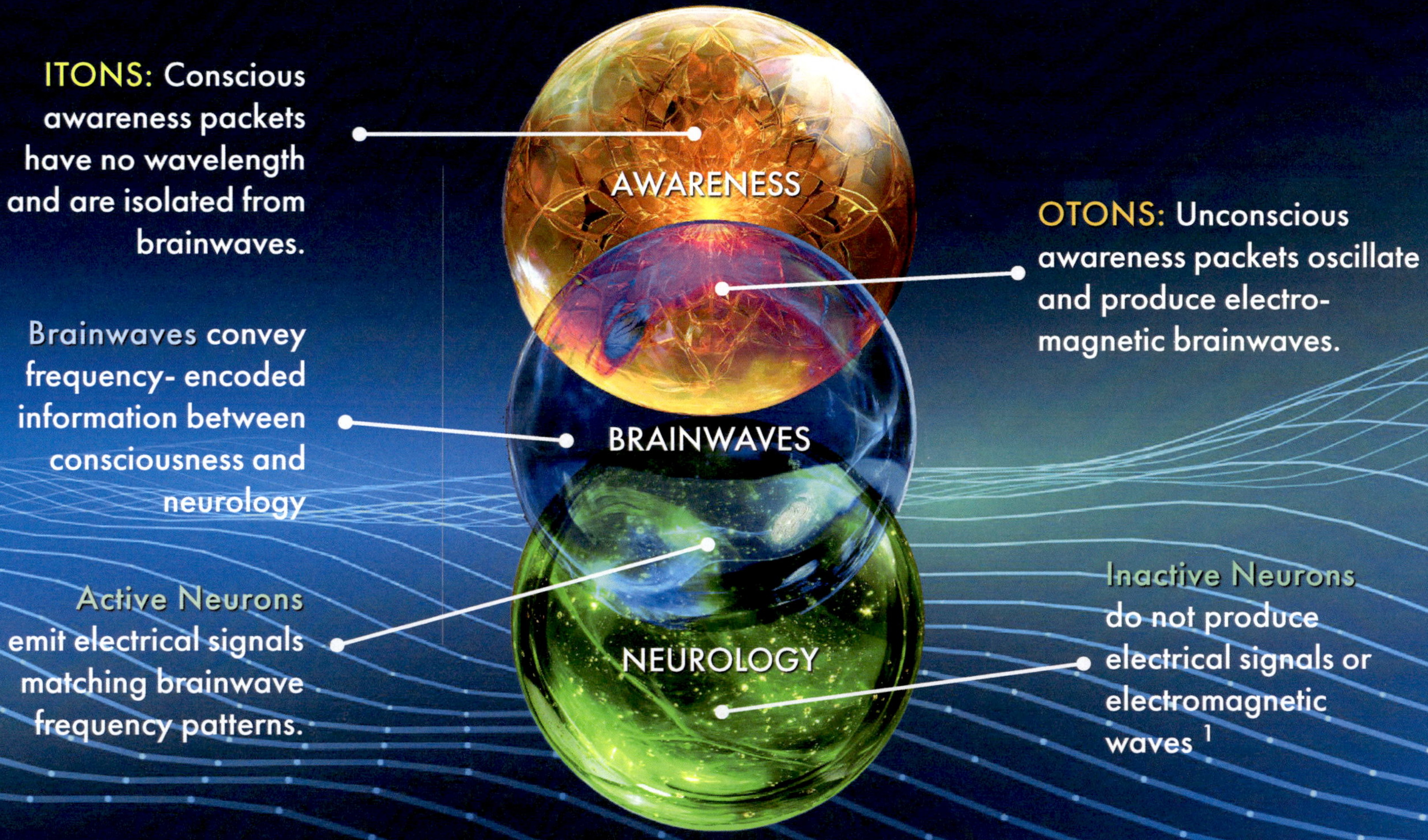

[1] Simplified diagram. The Reference Section contains a more nuanced treatment.

AXIOM 7

Conscious awareness varies inversely with neural activity and brainwave power.

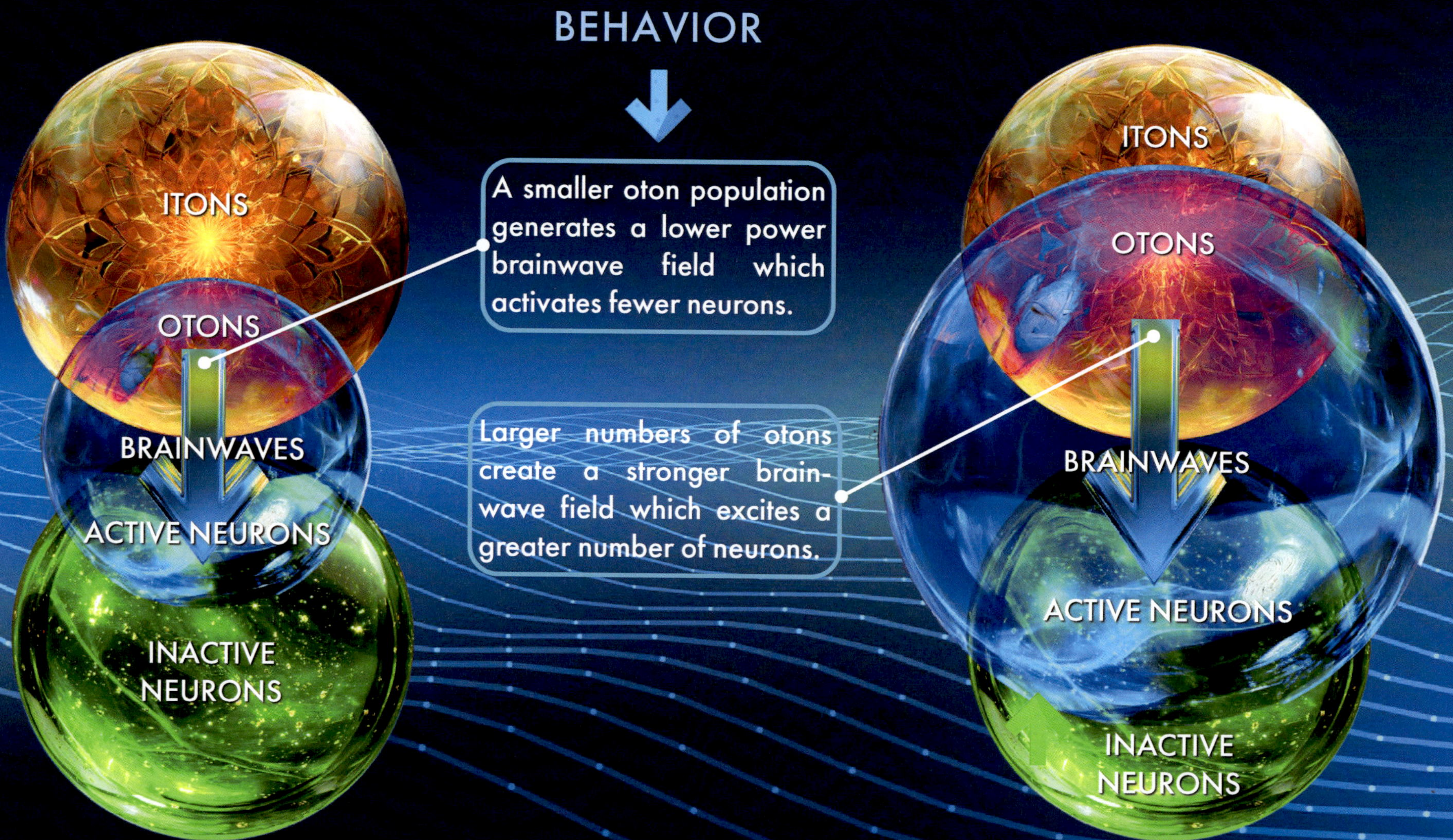

AXIOM 7

Conscious awareness varies inversely with neural activity and brainwave power.

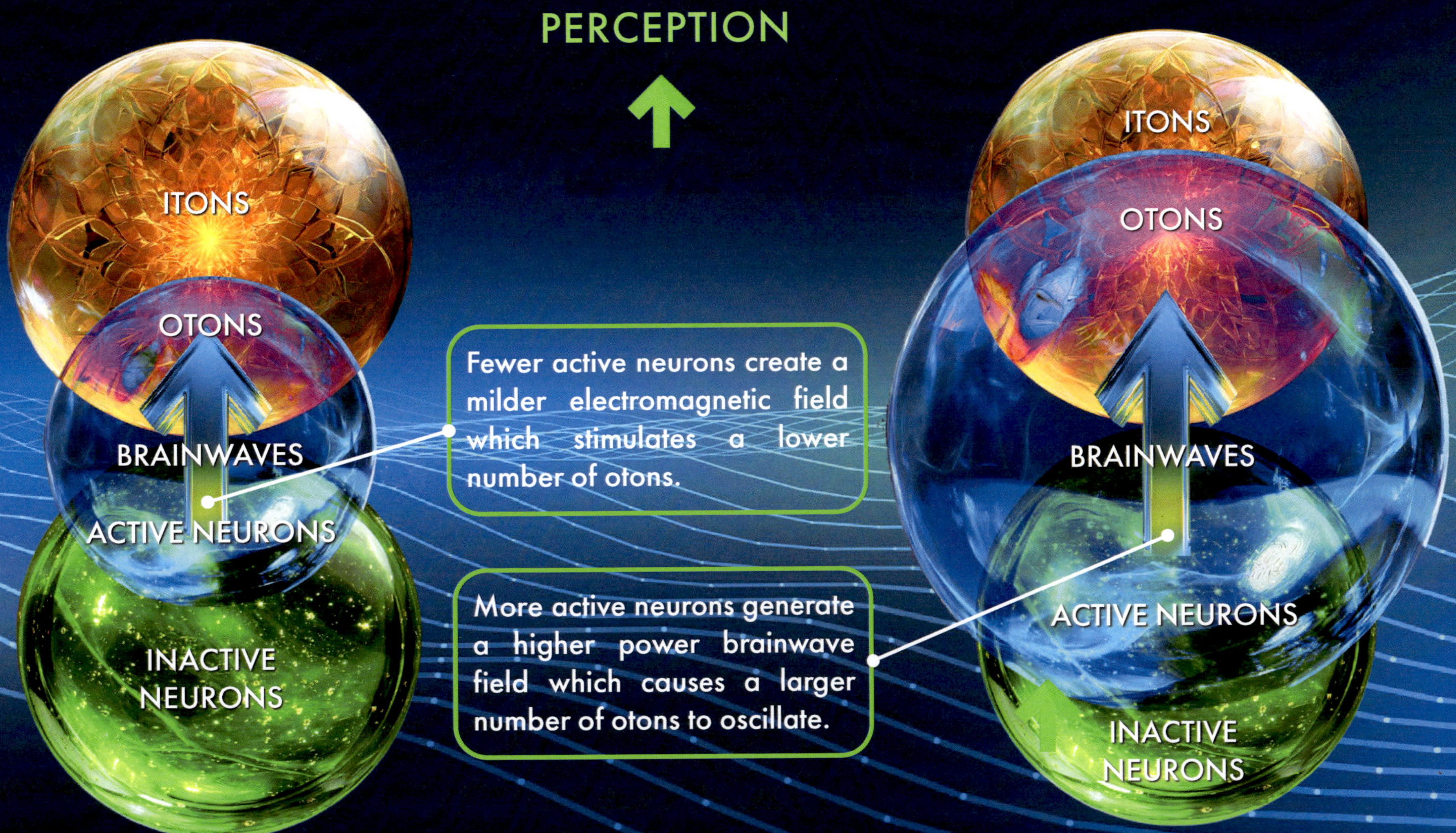

Neural Activity and Conscious Awareness – Converging Evidence

► Higher numbers of active neurons generate more current flow and power in neural networks.
► Axiom 7 posits an inverse relationship between conscious awareness and neural activity.
► Thus, conscious awareness should vary inversely with neural activity and the brain's power level.

This hypothesis is broadly supported in the scientific literature. A total evidence base of 112 peer-reviewed studies across 8 major research categories converges on a single principle:

Reduced excitatory neural activity correlates with enhanced conscious awareness and cognitive performance.

This section examines how this principle manifests in 6 major areas:

Neural efficiency in human intelligence

Default Mode Network deactivation during meditation and flow states

Multi-Matrix evidence summary

Brain power generation by cognitive state

Neurophysiological markers for meditation practitioners

Evidence strength by claim

Section 1: Neural Efficiency In Human Intelligence

1

PET and fMRI studies consistently demonstrate that individuals with higher cognitive ability show *less* neural activation when performing the same tasks as those with lower ability. This "neural efficiency" finding inverts the conventional assumption that more brain activity equals better thinking.

Table 1.1: Neural Efficiency by Performance Level

Performance Level	Neural Activation	Glucose Use	Relative Current
Expert / High IQ	Low	70–85% of baseline	↓ Reduced
Intermediate	Moderate	90–100% of baseline	— Baseline
Novice / Lower IQ	High	100–115% of baseline	↑ Elevated

Key Evidence

- Haier et al. (1988, 1992): High-IQ individuals use 15–30% less glucose during reasoning
- Neubauer & Fink (2009): Neural efficiency hypothesis—comprehensive review establishing that optimization ≠ maximization
- Dunst et al. (2014): Meta-analysis confirming neural efficiency across task types
- Grabner et al. (2006): Neural efficiency demonstrated in mathematical cognition
- *Nature Communications* study: Higher intelligence associated with lower dendritic density in cortical neurons; sparse, efficient neural networks outperform dense, redundant ones

Section 2: Brain Power Generation by Cognitive State

2

The brain's two primary networks draw between 10 and 25 watts of current depending on cognitive state.

Autonomic Network: The autonomic, homeostasis, metabolism, and maintenance functions in the brainstem, hypothalamus, insula, thalamus, and glial cells continuously draw ~10 watts of baseline current.

Cognitive Network: Memory and cognitive functions in the cortex and limbic areas raise the brain's total current to between 11 and 25 watts depending on activity level.

Table 2.1: Brain Power Generation by Cognitive State

Cognitive State	Power (W)	Evidence Level	Scientific Basis
Deep Meditation	10–12 W	Supported	Minimal DMN activity, high neural coherence; Newberg et al. (2001): 12% reduction in global cerebral blood flow; Lou et al. (1999) PET imaging; Brewer et al. (2011)
Light Meditation	13–16 W	Partial	10–20% reduction from baseline; Lazar et al. (2005); Tang et al. (2009)
Flow State	14–16 W	Partial	Reduced DMN activity, transient hypofrontality; Dietrich (2004); Ulrich et al. (2014) reduced medial prefrontal activity
Active Relaxation	~15 W	Partial	States like music listening less intensive than focus; Davidson et al. (2003); Benson et al. (1996): relaxation response reduces O_2 consumption 10–17%
Focused Attention	16–18 W	Supported	Prefrontal cortex activation, less than analytical reasoning; Posner & Petersen (1990); Fan et al. (2005)
Resting State	17–20 W	Confirmed	Well-established baseline; DMN remains active; Raichle & Gusnard (2002)
Analytical Mode	22–25 W	Supported	Complex problem-solving, logical reasoning; Haier et al. (1988); Peters, McEwen & Friston (2017) mapped stress metabolism up to 25W

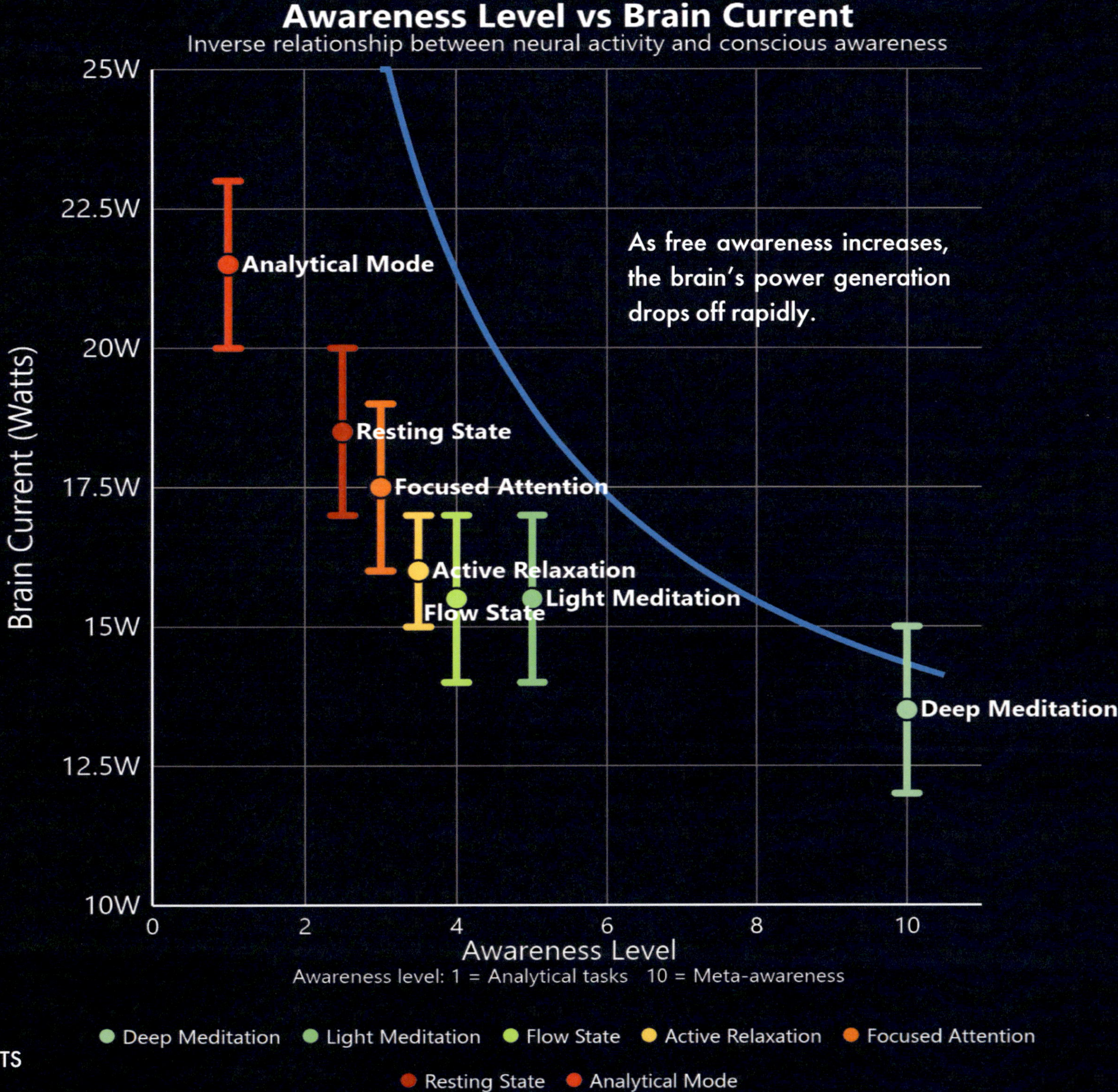
Awareness Level vs Brain Current
Inverse relationship between neural activity and conscious awareness
Brain Current (Watts)
25W
22.5W
20W
17.5W
15W
12.5W
10W
Analytical Mode
Resting State
Focused Attention
Active Relaxation
Flow State
Light Meditation
Deep Meditation
As free awareness increases, the brain's power generation drops off rapidly.
0
2
4
6
8
10
Awareness Level
Awareness level: 1 = Analytical tasks 10 = Meta-awareness
Deep Meditation
Light Meditation
Flow State
Active Relaxation
Focused Attention
Resting State
Analytical Mode

Table 2.2: Evidence Classification Summary

Category	States	Evidence Quality
Supported by Hard Scientific Data	Analytical Mode (22–25 W) Resting State (17–20 W) Deep Meditation (10–12 W)	Direct measurement
Extrapolations Based on Hard Data	Focused Attention (16–18 W) Flow State (14–16 W) Active Relaxation (~15 W) Light Meditation (13–16 W)	Derived from related studies

Core References for Brain Energy Metabolism

1. Raichle, M.E. & Gusnard, D.A. (2002). Appraising the brain's energy budget. *PNAS*, 99(16), 10237–10239.
2. Peters, A., McEwen, B.S. & Friston, K. (2017). How the selfish brain organizes its supply and demand. *Neuron*, 96(4), 742–754.
3. Clarke, D.D. & Sokoloff, L. (1999). Circulation and energy metabolism of the brain. *Basic Neurochemistry*, 6th ed.
4. Harris, J.J. & Attwell, D. (2012). The energetics of CNS white matter. *J. Neuroscience*, 32(1), 356–371.
5. Powers, R.E., Clark, T.M. & Ghuman, A.S. (2019). How much energy does the brain use? *BrainFacts.org*.

Section 3: Default Mode Network Deactivation

3

The Default Mode Network (DMN) activates during mind-wandering, self-referential thought, and rumination. Its attenuation is the most robust neural correlate of meditation, focused attention, and flow states. Less DMN activity consistently correlates with greater present-moment awareness.

Table 3.1: DMN Activity, Brainwave Frequency, and Awareness

Cognitive State	DMN Activity	Dominant Frequency	Relative Current	Awareness Level
Mind-Wandering	95%	Beta (13–30 Hz)	High	Low
Passive Rest	80%	Alpha-Beta (10–20 Hz)	Moderate-High	Baseline
Focused Attention	55%	Alpha (8–13 Hz)	Moderate	Elevated
Light Meditation	40%	Alpha-Theta (6–10 Hz)	Moderate-Low	High
Deep Meditation	20%	Theta (4–8 Hz)	Low*	Very High
Flow State	15%	Theta-Alpha (5–10 Hz)	Low*	Peak

Notes and Key References

1. The primary DMN regions are the posterior cingulate cortex and medial prefrontal cortex.
2. Brain current flow varies inversely with brainwave frequency per capacitive reactance: $X_c = 1/(2\pi fC)$

- Brewer et al. (2011): Meditators show 40–60% DMN reduction — *PNAS*
- Garrison et al. (2015): Real-time fMRI confirms DMN-awareness relationship
- Raichle (2015): The brain's default mode network — *Annual Review of Neuroscience*

Awareness Level vs Brain Power Generation

DMN Suppression as empirical proxy for free awareness (itons)

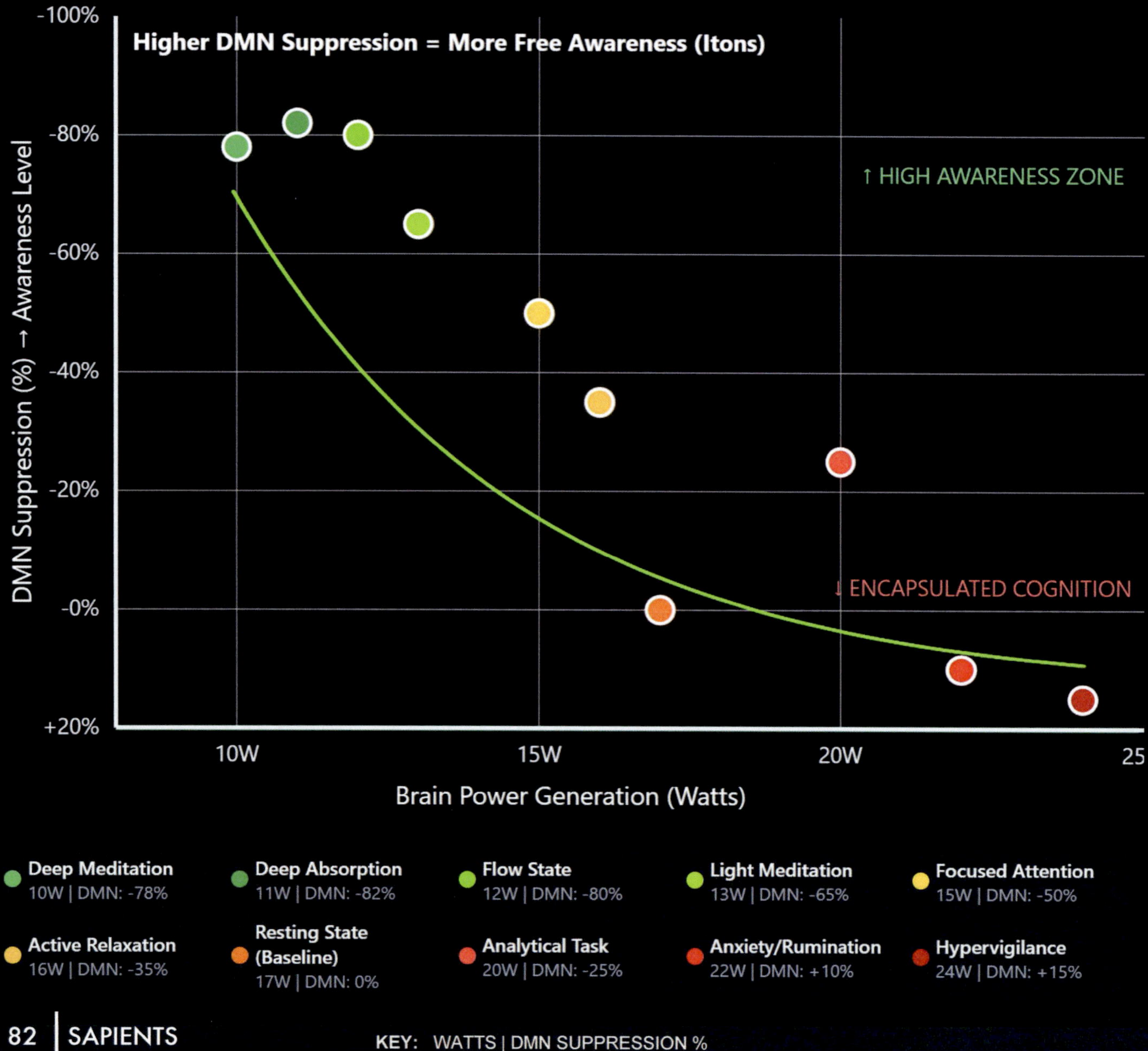

DMN Suppression — Th
Default Mode Network (DMN) i
active during self-referential
thinking, mind-wandering, and
rumination—all processes that
consume awareness packets (o
DMN suppression correlates wi

- Less self-referential processi
- More present time awarene
- Enhanced metacognitive cla
- Lower metabolic activity in medial prefrontal and poste cingulate cortices

Sources: Brewer et al., 2011 (PN Garrison et al., 2015; Ulrich et a 2014; Carhart-Harris et al., 2012

Key Insight: The hyperbolic curve demonstrates why sma reductions in brain power (vi 5-HT2A modulation) produc large gains in awareness— consistent with Corollary 6 and Ohm's Law.

Section 4: Neurophysiological markers for meditation practitioners

4

Multiple neurophysiological and behavioral metrics change with meditation experience. The convergence of these independent measures reinforces evidence for the underlying relationship between reduced neural excitation and enhanced awareness.

Table 4.1: Neurophysiological Changes with Meditation Experience

Metric	Novice (0 hrs)	→ Expert (10,000 hrs)	Change	Current Flow
DMN Activity	High	→ Low	↓ 40–60%	↓ Reduced
Beta Power	High	→ Low	↓ 20–35%	↓ Reduced
Neural Noise	High	→ Low	↓ 30–50%	↓ Reduced
Firing Rate (5-HT2A KD)	Baseline	→ Reduced	↓ 24%	↓ Reduced
Alpha Coherence	Low	→ High	↑ 25–40%	— Stable*
Theta Power	Low	→ High	↑ 30–50%	↓ Reduced†
Attention Score	Baseline	→ Elevated	↑ 15–30%	— (behavioral)
Reported Clarity	Baseline	→ High	↑ Significant	— (subjective)

Key References

Cahn & Polich (2006): Meditation states and traits — *Psychological Bulletin*

Lutz et al. (2004): Long-term meditators show gamma synchrony — *PNAS*

Tang et al. (2015): Neuroscience of mindfulness meditation — *Nature Reviews Neuroscience*

Brefczynski-Lewis et al. (2007): Neural correlates of attentional expertise in long-term meditators — *PNAS*

Directional Summary

↓ *Decreases with Experience:* DMN activity, Beta power, Neural noise, Firing rate, Net current flow

↑ *Increases with Experience:* Alpha coherence, Theta power, Attention scores, Reported clarity

Notes

Alpha coherence increases without a proportional current increase due to improved synchronization efficiency.

†Theta power increases but at lower frequency, resulting in net current reduction by raising capacitive reactance $Xc = 1/(2\pi fC)$.

Section 5: Multi-Metric Evidence Summary

5

The relationship between reduced neural excitation and enhanced awareness is supported by multiple independent research programs across different methodologies, populations, and measurement techniques.

Table 5.1: Converging Evidence Across Research Domains

Domain	Relationship	Current Effect	Key Finding	Key Studies
DMN Activity	Inverse ↓↑	↓ Current	40–60% reduction in meditators	Brewer et al. (2011); Garrison et al. (2015)
Neural Efficiency	Inverse ↓↑	↓ Current	15–30% less glucose use in high-IQ individuals	Haier et al. 1992); Neubauer & Fink (2009)
EEG Beta Power	Inverse ↓↑	↓ Current	Excess beta correlates with anxiety, rumination	Cahn & Polich (2006)
5-HT2A Expression	Inverse ↓↑	↓ Current	Knockdown = 94% memory improvement	Rohn et al. (2024) *Translational Psychiatry*
Neuron Firing Rate	Inverse ↓↑	↓ Current	5-HT2A knockdown = 24% firing rate reduction	Rohn et al. (2024)
Anxiety	Inverse ↓↑	↓ Current	5-HT2A knockdown = 34% reduction	Rohn et al. (2024); Weisstaub et al. (2006)
Hippocampus	Inverse ↓↑	↓ Current	Hyperactivity impairs memory; reducing it restores	Bakker et al. (2012) *Neuron*
Depression	Inverse ↓↑	↓ Current	10–20% binding reduction correlates with improvement	Bhagwagar et al. (2006); Meyer et al. (2001)
Longevity	Inverse ↓↑	↓ Current	Centenarians have "quieter" brains; REST elevated	Zullo et al. (2019) *Nature*; Lu et al. (2014)
Alpha Coherence	Direct ↑↑	— Neutral	Meditation raises alpha synchronization	Lutz et al. (2004)
Theta Power	Direct ↑↑	↓ Current	Deep meditation shows elevated frontal theta	Tang et al. (2015)
Sparse Coding	Direct ↑↑	↓ Current	Optimal information requires minimal redundancy	Genç et al. (2018) *Nature Communications*
Attention	Inverse ↓↑	↓ Current	5-HT2A agonists impair; antagonists improve	Quednow; M100907 studies
Total Brain Metabolism	Inverse ↓↑	↓ Current	Global changes modest; regional shifts larger	Raichle (2006); Shulman et al. (2004)

Section 6: Evidence Strength by Claim

6

112 Studies Across 12 Categories

Table 6.1: Evidence Base Classification

Claim	Evidence Level	Key Finding
5-HT2A↓ reduces anxiety	Very Strong	34% reduction with knockdown
5-HT2A↓ improves memory	Very Strong	94% improvement in object recognition
Hippocampal hyperactivity → memory impairment	Very Strong	Bakker et al. (2012)
Neural excitation accelerates aging	Very Strong	Zullo et al. (2019) *Nature*, cross-species
REST suppression of excitation extends lifespan	Very Strong	Causal manipulation, conserved across species
5-HT2A↓ reduces depression markers	Strong	PET shows elevated binding in depression
5-HT2A↓ improves attention	Strong	Direct antagonist studies show improvement
5-HT2A↓ improves cognition	Strong	Multiple domains improved
Neural efficiency (↓density → ↑IQ)	Strong	*Nature Communications* dendritic study
Long-term safety supported	Strong	CATIE 18-month, 10-year studies
5-HT2A↓ reduces hippocampal hyperactivity	Moderate-Strong	Convergent evidence supports pathway
5-HT2A↓ reduces limbic hyperactivity	Moderate-Strong	Treatment normalizes amygdala responses

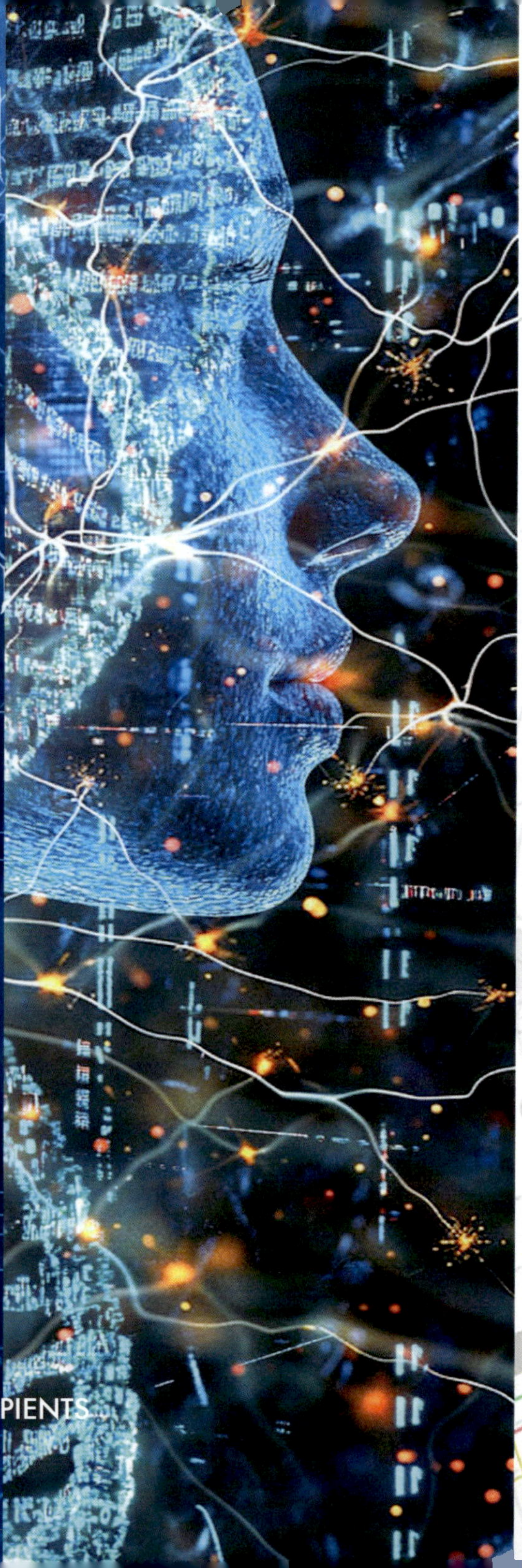

Strategy

01

Axiom 7: Conscious awareness varies inversely with neural activity and brainwave power.

02

Evidence: Reduced excitatory neural activity correlates with enhanced conscious awareness and cognitive performance.

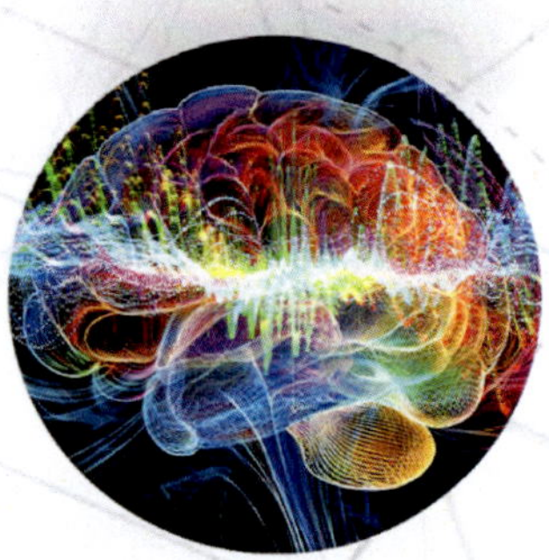

03

Design: Neural activity and current flow can be decreased by genetically lowering unnecessary neuron signaling activity.

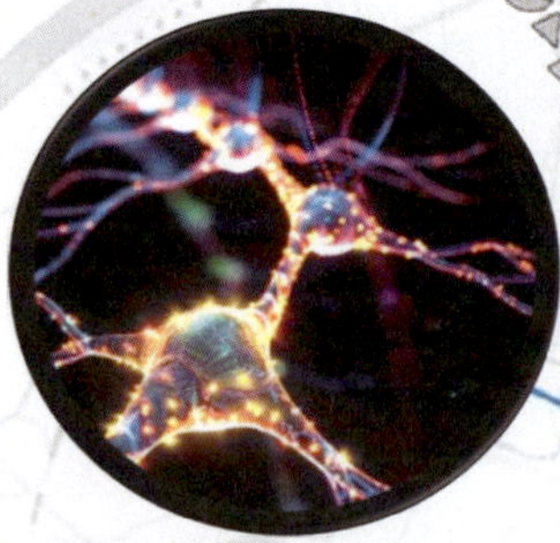

Definitions

Electronics

Resistor: A circuit element which offers opposition to current.

Resistance: The opposition a resistor offers to current.

Capacitor: A circuit element which stores and accumulates electric charge and releases it when it reaches a threshold.

Capacitance: The amount of charge a capacitor can store and hold (a constant).

Capacitive Reactance: The opposition a capacitor offers to current, which decreases as the frequency of the voltage increases.

RC Circuit: An electrical circuit with resistors and capacitors.

Impedance: An RC circuit's total opposition to current, combining its resistance, capacitance and capacitive reactance.

Neurology

Neuron as a resistor: A neuron selectively admits charged particles based on its activation level and membrane permeability. Its activation level depends on the number of open channels it has for receiving activation signals.

The channels are called neurotransmitter receptors. More open receptors offer more doorways into the neuron, reducing its resistance and allowing more current to flow through the network.

Neuron as a capacitor: A neuron receives, accumulates and stores electrically charged particles, releasing them when the charge reaches a threshold level.

RC Circuit: The neuron's electrical model is a resistor-capacitor (RC) circuit.

Receptors

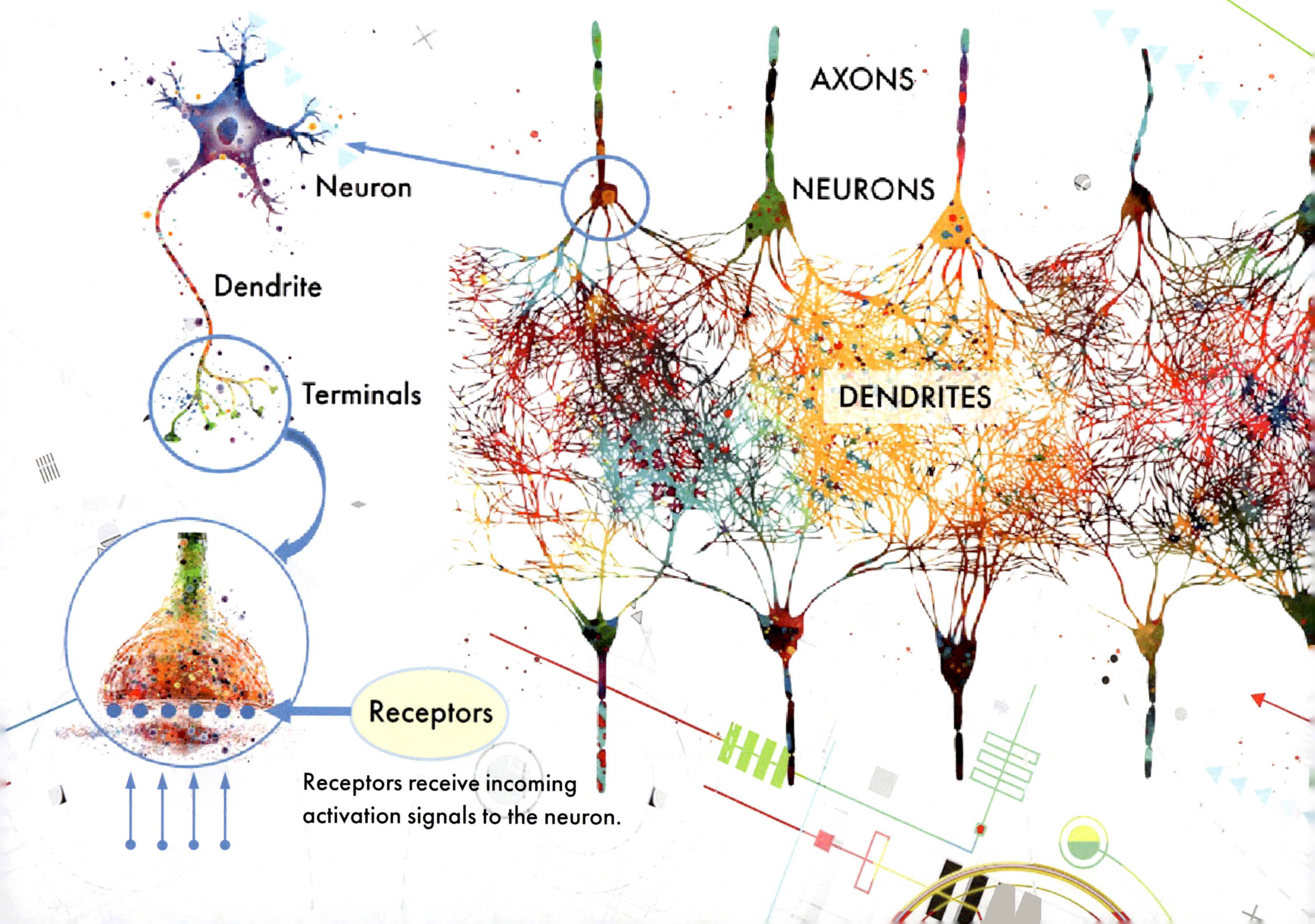
AXONS
Neuron
NEURONS
Dendrite
Terminals
DENDRITES
Receptors
Receptors receive incoming activation signals to the neuron.

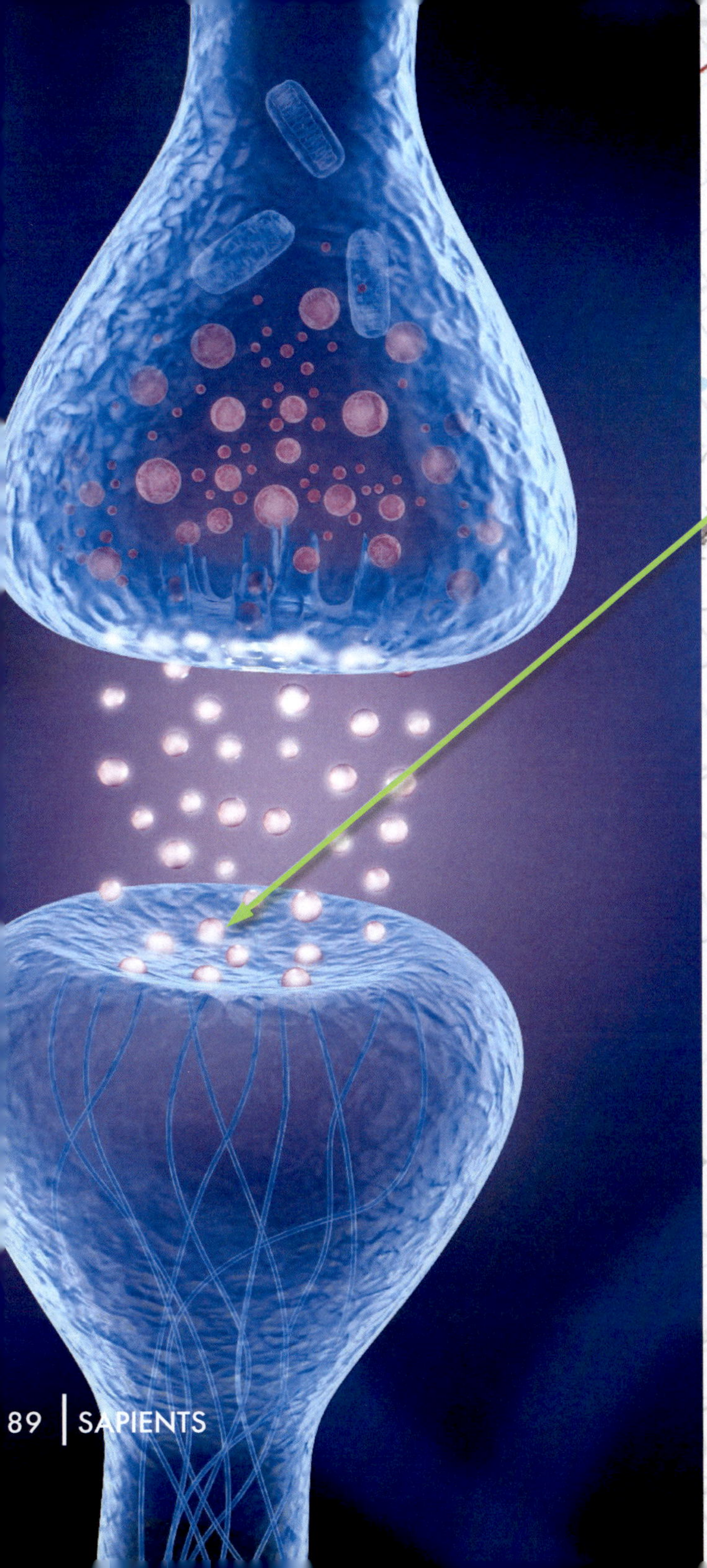

Neurotransmitters

The brain sends messages between neurons using 4 chemical messengers called neurotransmitters shown below.

Neurotransmitters work by stimulating receptors which act as doorways into neurons for receiving incoming signals. These signals may excite or inhibit their activity.

Serotonin

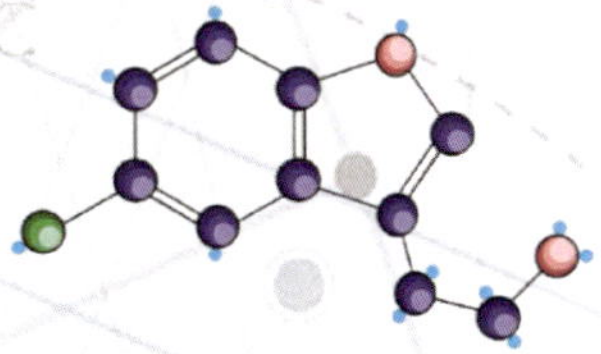

Regulates mood, cognition and emotion

Glutamate

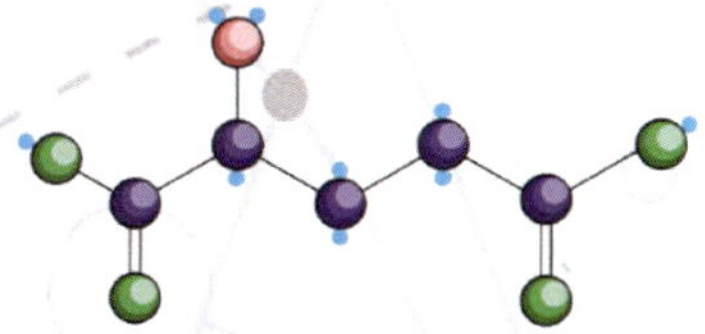

Excites neurons and activates brain circuits

Dopamine

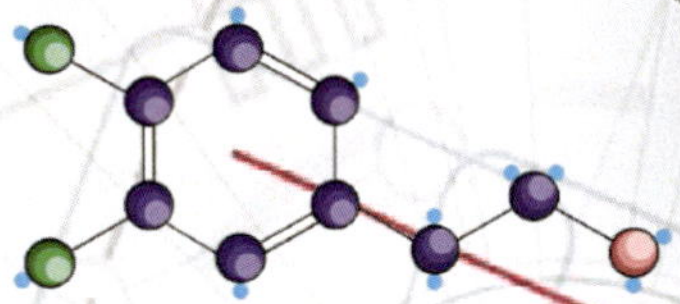

Regulates motivation, reward and movement

Gaba (gama amino-butyric acid.)

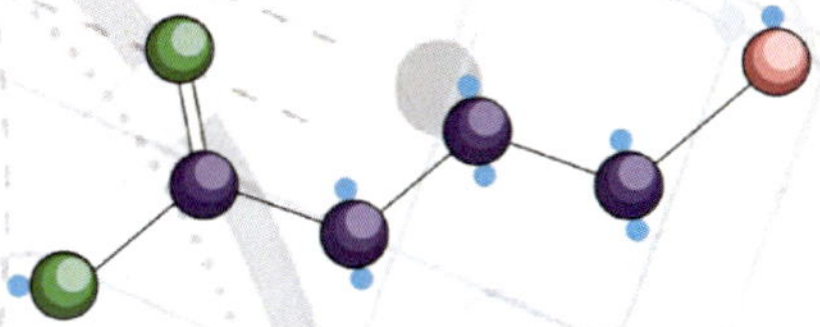

Inhibits neurons and reduces brain activity

The Unlock

There are 14 subtypes of receptors for serotonin. One type, the serotonin 2A receptor (5-HT2A), has far-reaching effects on cognition.

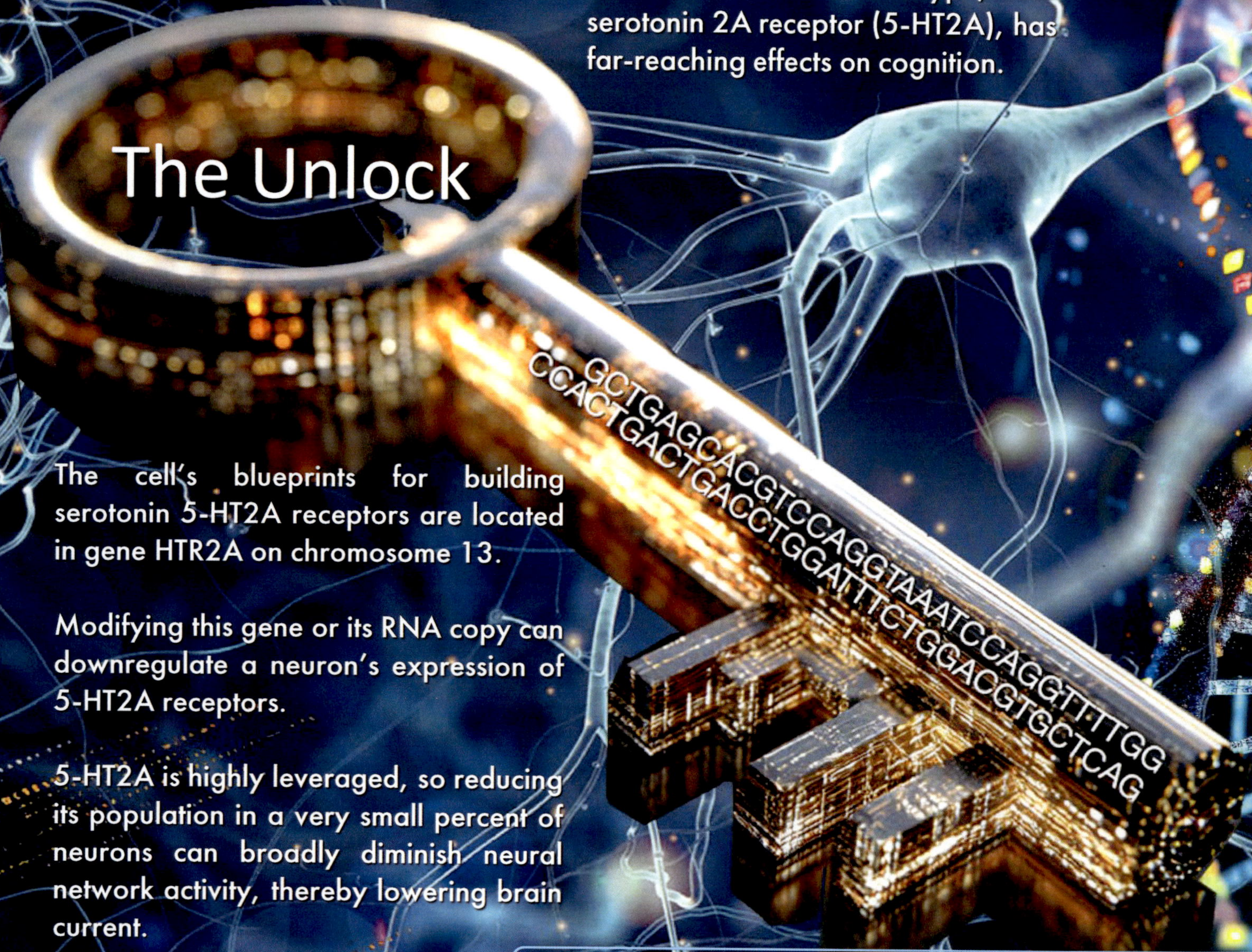

The cell's blueprints for building serotonin 5-HT2A receptors are located in gene HTR2A on chromosome 13.

Modifying this gene or its RNA copy can downregulate a neuron's expression of 5-HT2A receptors.

5-HT2A is highly leveraged, so reducing its population in a very small percent of neurons can broadly diminish neural network activity, thereby lowering brain current.

In electrophysiology studies with cortical neuron cultures, knocking out 5-HT2A receptors lowered neuronal firing rates rate by 24%. (41)

90 | SAPIENTS

Mechanisms of Action

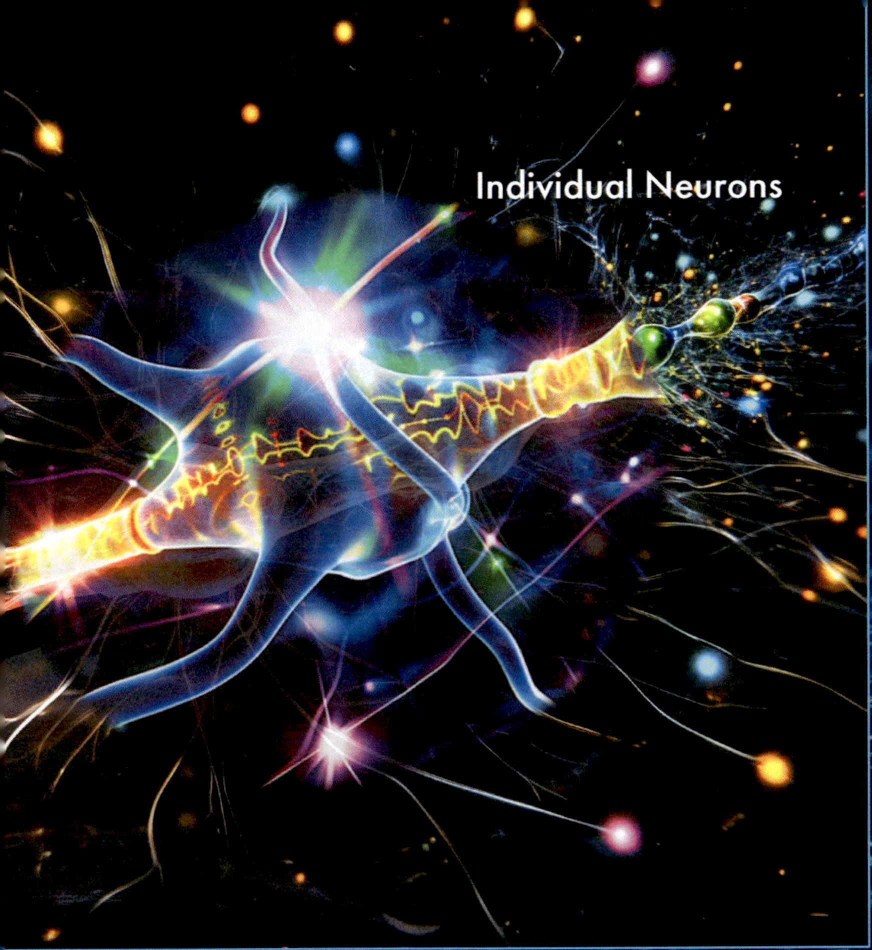

METHOD 1: ELECTRICAL IMPEDANCE

A neuron is like a pump on a water line. If the pump slows down, less water flows through the line. If a neuron fires less often, less current flows through the network.

METHOD 2: CHEMICAL SIGNALING

Reducing one neuron's 5-HT2A receptor activity can dampen glutamate release in as many as 600 downstream neurons. Since glutamate is the primary excitatory neurotransmitter, lowering its concentration attenuates neuron activity globally.

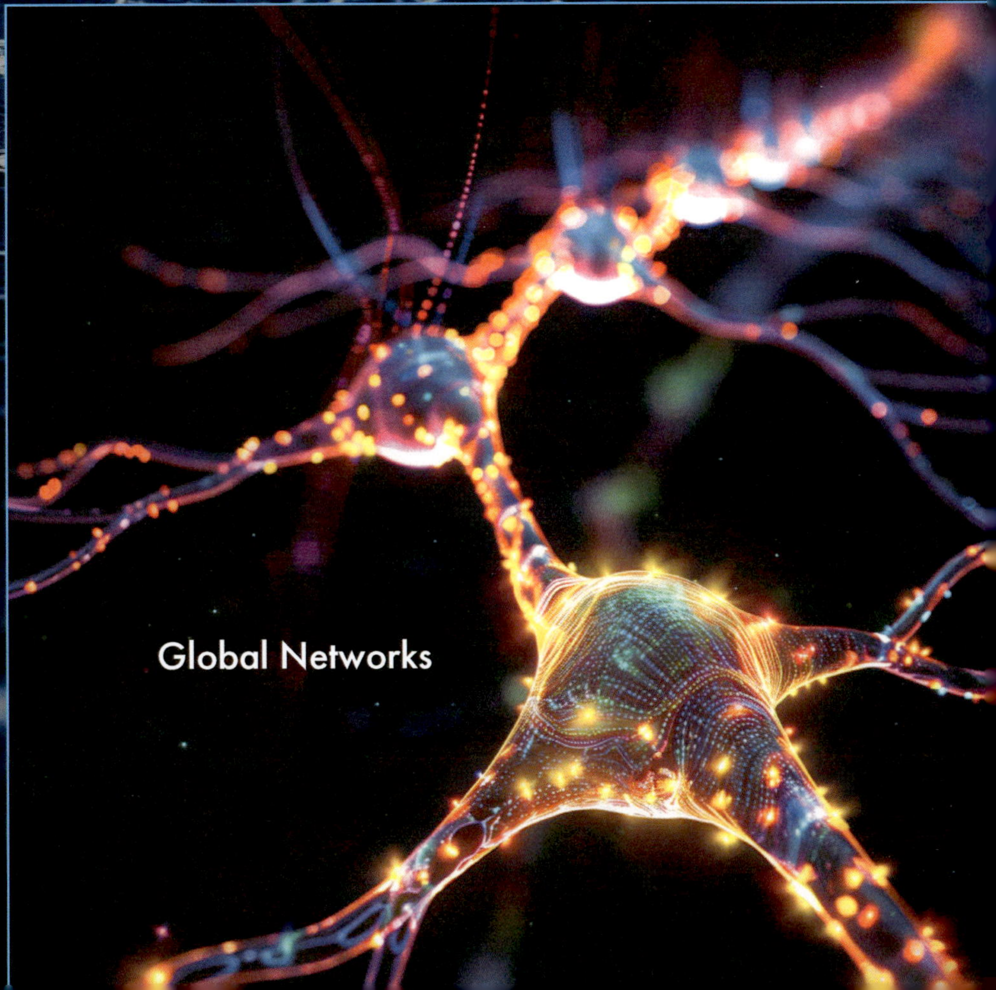

Methodology Summary

Goal: Decrease current flow in the brain's cortical and limbic networks

Local Networks Strategy

Raise neural resistance and reduce neuron signaling rate in targeted neurons.

Global Activity Strategy

Raise neural impedance (1) and reduce glutamate levels in downstream neurons.

1 Receptor Density

Downregulating 5-HT2A receptor population raises a neuron's resistance and its opposition to incoming signals.

2 Signaling Rate

Modified neurons charge more slowly and fire less often, lowering signaling frequency.

3 Reduced Current

Lower frequency neurowave voltage reduces current in neural networks.

4 Higher Reactance

Lower frequency voltage increases capacitive reactance (2) in downstream neurons.

5 Glutamate Levels

Reducing 5-HT2A receptor activity dampens glutamate release in downstream neurons.

6 Global Activi

Decreased glutamate level attenuate neuro activity globally

(1) Impedance = [resistance + capacitance + capacitive reactance]

(2) Capacitive reactance = the opposition a capacitor offers to current

(3) Glutamate is the primary excitatory neurotransmitter.

Neuron Electrodynamics

Neural resistance drives frequency and current.

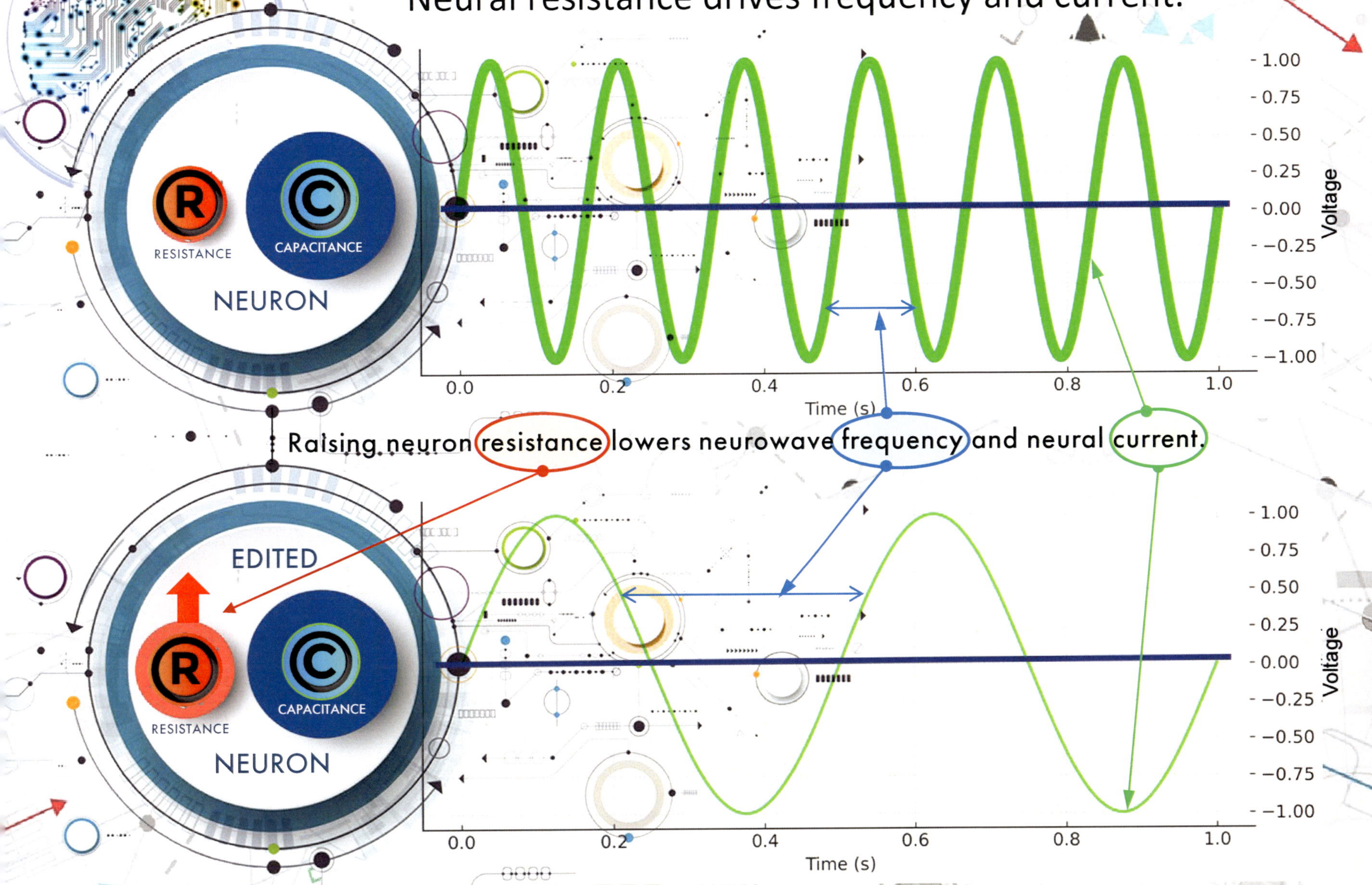

Note: This relationship applies to excitatory receptors , i.e. certain serotonin, glutamate and dopamine subtypes.

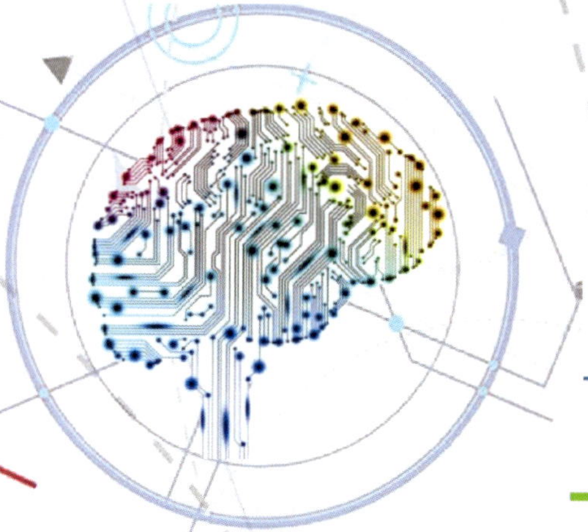

Neurowave Frequency

1. Neural resistance drives neurowave frequency.

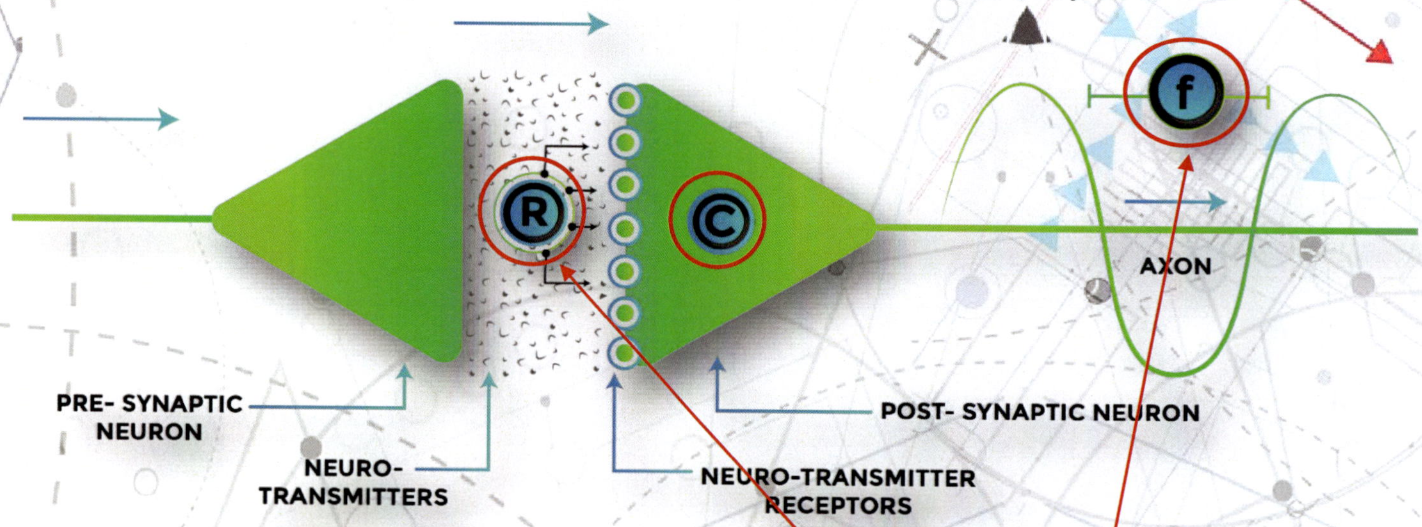

The relationship between neurowave frequency and neuron resistance can be expressed as:

$$f = \frac{1}{2\pi RC}$$

where:

- Frequency f is the number of neuron pulses per second or the wavelength between peaks.
- Resistance R varies with the number of neurotransmitter receptors receiving activation signals.
- Capacitance C is the neuron's ability for storing electrical charge. It is basically a constant.

A neuron's pulse frequency f varies with its electrical resistance R.

Thus, resistance R determines frequency f.

Note: This equation applies to excitatory receptors , i.e., certain serotonin, glutamate and dopamine subtypes.

Neural Current

2. Neurowave frequency drives neural current.

Neural current varies in proportion to neural voltage frequency.
Higher frequencies allow more current to flow.
Lower frequencies reduce current flow.

Current-Frequency Relationship
in Neural Resistor/Capacitor (RC) Circuits

$$X_C = 1 / (2 \pi f C)$$

Where:

X_C = capacitive reactance (the opposition a capacitor offers to current)
f = frequency
C = capacitance (the amount of charge a capacitor can hold) (a constant)
Neurons function electrically like resistor/capacitors, accumulating and releasing charge.

As frequency decreases, capacitive reactance increases, reducing current flow. Theta frequencies (4-8 Hz) produce higher reactance and lower current than Beta frequencies (13-30 Hz).

Genetic Control

3. Neural resistance can be genetically modified.

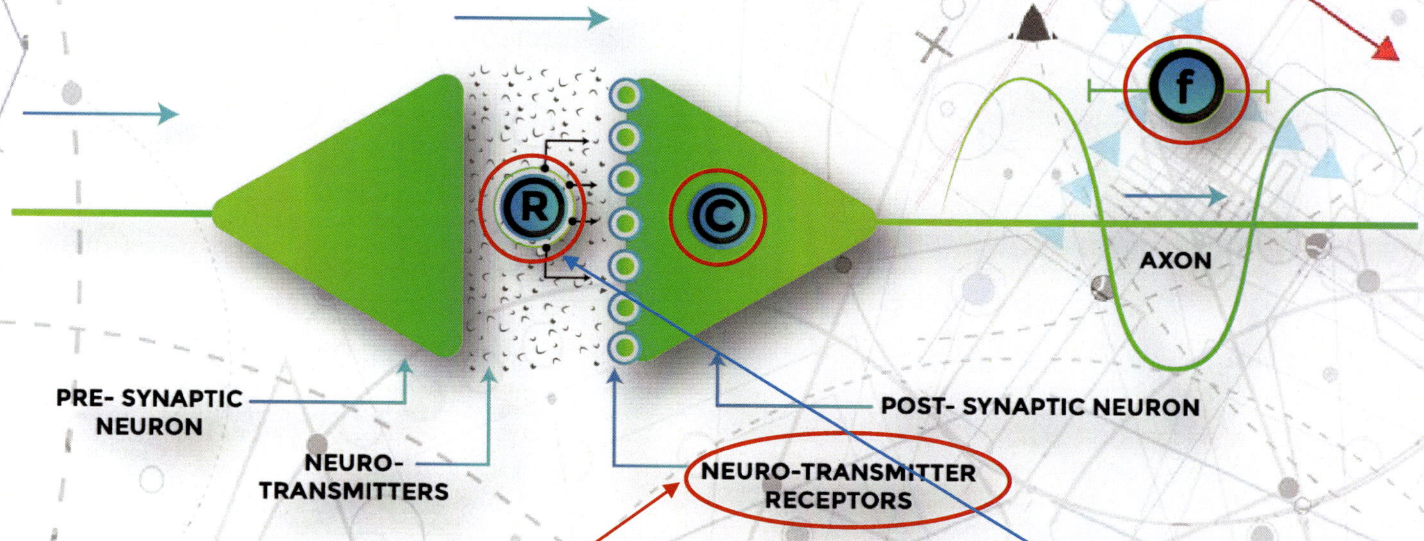

STEP 1 Neurons receive excitatory signals through gateways called neuro-transmitter receptors.

STEP 2 A neuron's electrical resistance R is largely determined by the number of these gateways open for signals to enter the neuron.

STEP 3 A neuron's resistance R can be increased by reducing its neuro-transmitter receptor population.

STEP 4 A neuron's receptor population can be controlled by modifying its DNA or RNA.

Methodology Schematic

Neural current attenuation strategies

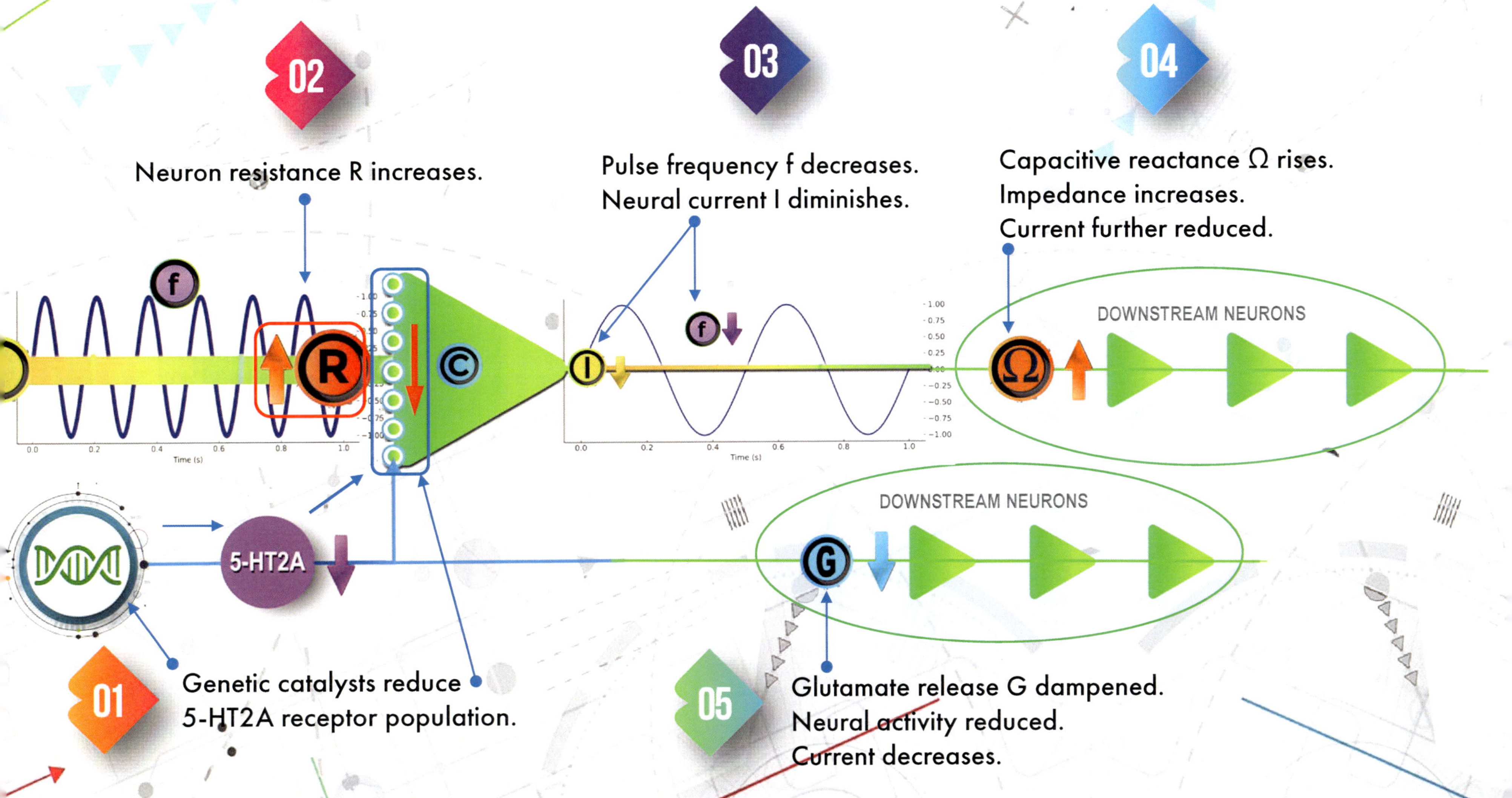

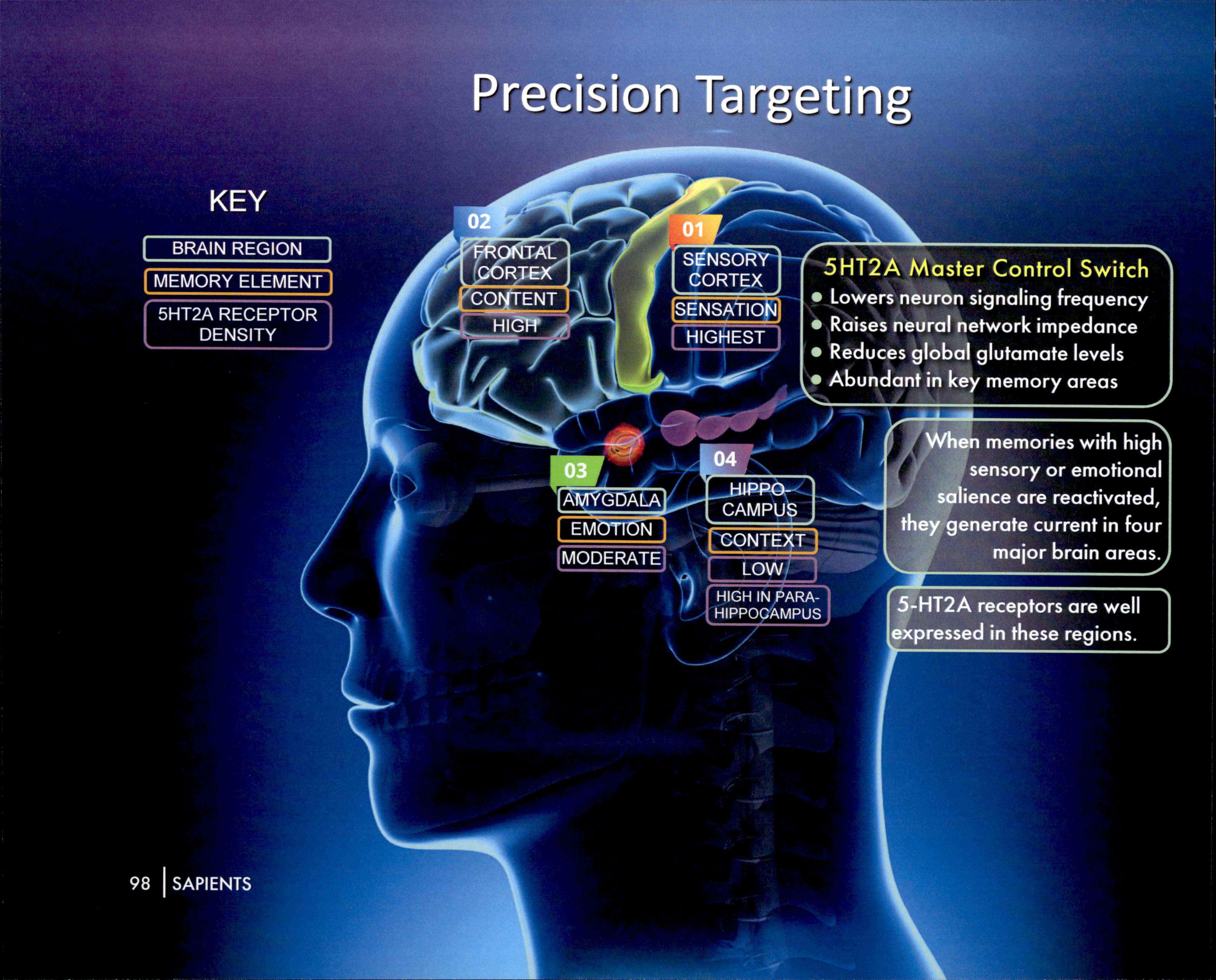
Precision Targeting
KEY
BRAIN REGION
MEMORY ELEMENT
5HT2A RECEPTOR DENSITY
02
FRONTAL CORTEX
CONTENT
HIGH
01
SENSORY CORTEX
SENSATION
HIGHEST
03
AMYGDALA
EMOTION
MODERATE
04
HIPPO-CAMPUS
CONTEXT
LOW
HIGH IN PARA-HIPPOCAMPUS
5HT2A Master Control Switch
• Lowers neuron signaling frequency
• Raises neural network impedance
• Reduces global glutamate levels
• Abundant in key memory areas
When memories with high sensory or emotional salience are reactivated, they generate current in four major brain areas.
5-HT2A receptors are well expressed in these regions.

Optimization

01 Limbic memory circuits typically carry higher current than cortical circuits, especially for memories with emotional content.

02 Editing limbic neurons can impede smore current than cortical neurons on a one for one basis.

03 Genetic catalysts can be preferentially targeted to the limbic region via intranasal delivery.

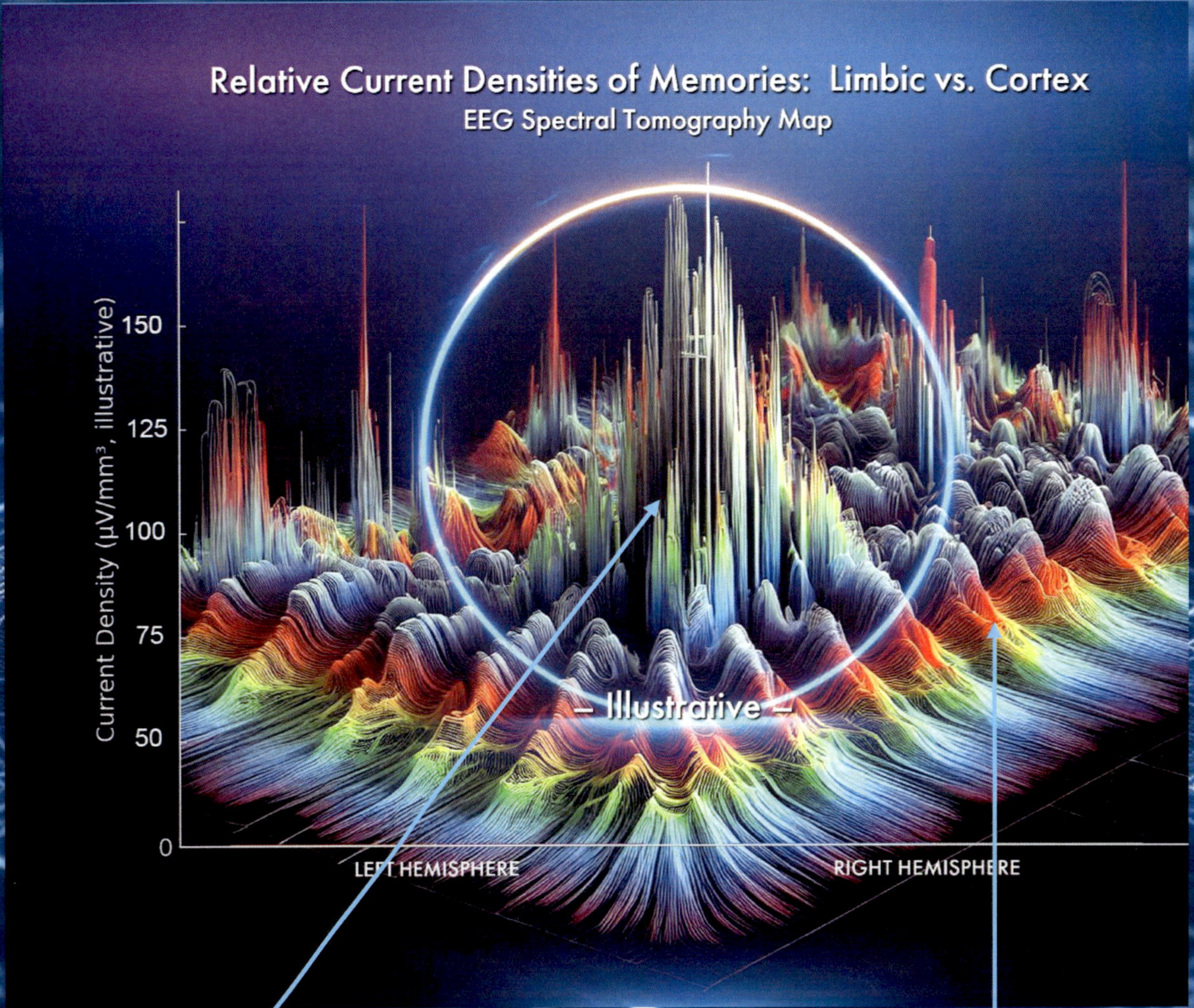

Limbic regions* generate 2–3 times higher field amplitudes than the cortex during memory encoding, due to synchronized sharp-wave ripple bursts.

The cortex distributes lower voltage signals broadly across large networks for sensory detail and integration.

* e.g., hippocampus/amygdala

Safety

Experimental evidence confirms that down-regulating 5-HT2A improves cognition in animal models without disrupting autonomic and metabolic functions.

- 5-HT2A is not expressed in the brainstem or hypothalamus. These regions maintain autonomic stability independent of 5-HT2A.

- Autonomic functions rely on other serotonin receptors (5-HT1A, 5-HT2C, 5-HT3), not 5-HT2A.

- Metabolism is regulated by multiple neurotransmitter systems, with minimal 5-HT2A involvement.

Conclusions

All cognitive network activity is information processing. The Sapient HTR2A edits will begin to shift individuals out of compulsively animating inessential unconscious information; restoring their natural awareness and sovereignty.

This change will consolidate more free awareness in present time, increasing the person's reservoir of cognitive capital. Sapients can consciously apply this capital to address any goal, task or problem with greater agency.

Risks & Issues

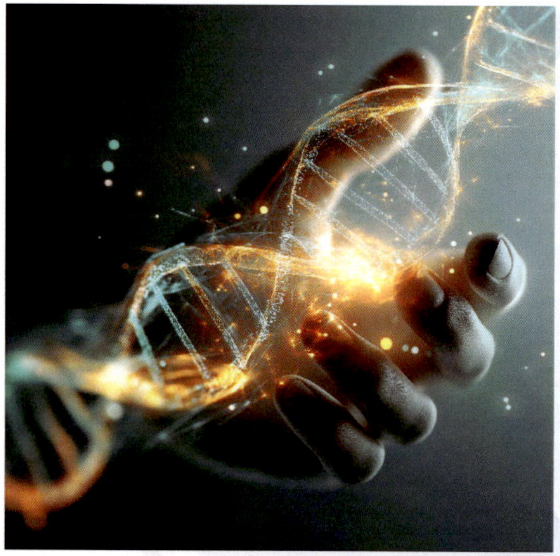

1. The New Cognitive Divide
Stratification between enhanced and unenhanced populations could create a permanent underclass. Access may be determined by wealth, geography, or political favor. Intergenerational compounding will occur as enhanced parents produce advantages for children, while unenhanced families fall further behind.

2 Cognitive Capital Distribution
Enhanced individuals could accumulate disproportionate wealth and influence. Investment returns will favor the cognitively enhanced. New monopolistic dynamics may emerge as enhanced founders and executives dominate markets.

3. Global Cognitive Geopolitics
Nations will compete to enhance military, intelligence, and leadership personnel. First-mover advantages will create new power asymmetries. Cognitive enhancement risks becoming classified military technology, restricting civilian access while accelerating covert programs.

4. Labor Market Disruption Enhanced workers may displace unenhanced workers in knowledge economies, compounding AI displacement. Wage compression will affect cognitive tasks. Credentialing crises will emerge as traditional qualifications become less meaningful against verified cognitive enhancement.

5. Identity and Psychological Destabilization Individuals face risks of personality discontinuity, relationship issues, and existential disorientation. Depression or dissociation may follow enhancement. Enhanced individuals may lose connection with unenhanced family members and communities.

6. Regulatory Arbitrage and Black Markets Enhancement tourism will emerge to unregulated jurisdictions. Underground providers will offer unverified treatments. Coerced enhancement may occur in employment or military contexts. Verification and enforcement will pose significant challenges

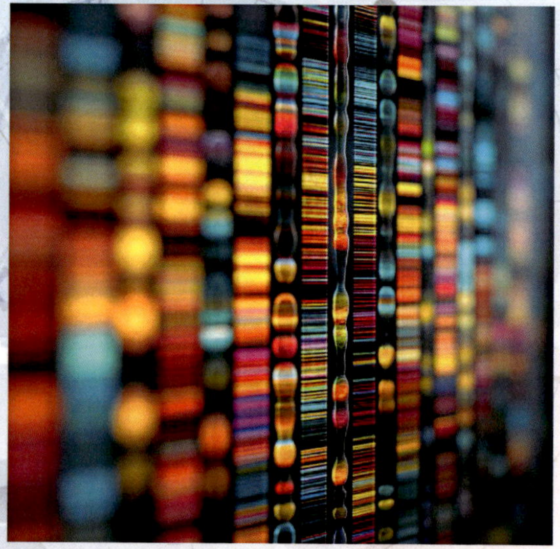

7. Community Fragmentation
Social communities may divide along enhanced and unenhanced lines. Religious and philosophical conflicts will arise over human identity. Democratic legitimacy questions will surface when enhanced minorities hold disproportionate power.

8. Corporate Cognitive Control
Proprietary cognitive technologies could create dependency relationships. Cognitive data harvesting will erode privacy. Enhancement may be tied to employment contracts or subscription models. Ongoing biological access creates manipulation risks.

9. The Homogenization Paradox
Enhancement toward uniform cognitive profiles may reduce the creative diversity that drives innovation. Groupthink could emerge at scale among similarly-enhanced leadership classes. Society may lose perspectives, intuitions, and cognitive styles that arise from unoptimized neurology.

De-Risking Strategy

The Sapient Initiative outlines key de-risking provisions but further measures are necessary.

<table>
<tr><th>New Cognitive Divide</th><th>Cognitive Capital Distribution</th><th>Global Cognitive Geopolitics</th><th>Labor Market Disruption</th><th>Identity and Psychological Destabilization</th></tr>
<tr><td>Mitigated through tokenization and scholarship fund to democratize access.</td><td colspan="3">Require additional policy frameworks beyond the initiative's current scope.</td><td>Addressed through development of therapeutic protocols and Level 100's curriculum designed to prime candidates psychologically for change.</td></tr>
</table>

Regulatory Arbitrage and Black Markets	Community Fragmentation	Corporate Cognitive Control	Homogenization Paradox
Countered by robust policy development governing research ethics and clinical applications. The blockchain registry also provides auditability and quality control to distinguish legitimate providers from underground operators.	Requires additional policy frameworks	Partially addressed through the blockchain's transparency mechanisms and the token model's decentralized ownership structure.	Requires additional policy frameworks

Chapter 9: Future Directions

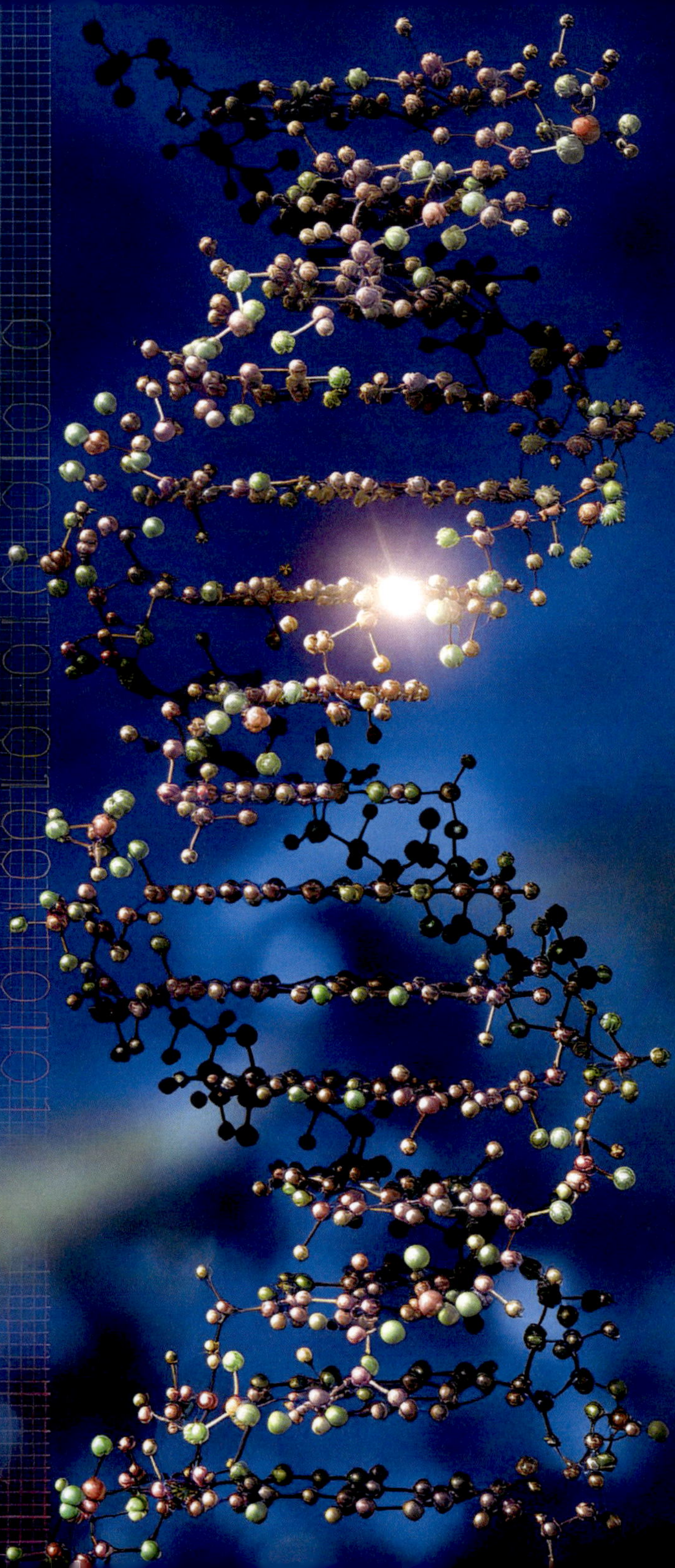

9.1: Scientific evolution: Future neuroimaging instruments and research directions

9.2: Ethical use: Frameworks for responsible innovation in cognitive genetics technologies

9.3: Further exploration: Studies inviting rigorous inquiry, critique and collaboration

9.4: New awareness science: Consciousness as a primary domain embracing psi phenomena

– Mathematical models of awareness and comparisons with 200 consciousness theories

– Memory theory correlating active and inert memories with neural current flow

– Integration of memory, EM field dynamics, and awareness into a unified framework

9.1 Scientific Evolution

The axioms and formulas in this book form the kernel of a new awareness science based on information from classical neuroimaging instruments such as EEG, PET and fMRI.

Next-generation systems, such as magnetic encephalography (MEG), virtual neuroimaging, and MEG-EEG hybrids, promise to deliver new levels of precision that will allow researchers to refine and expand the science.

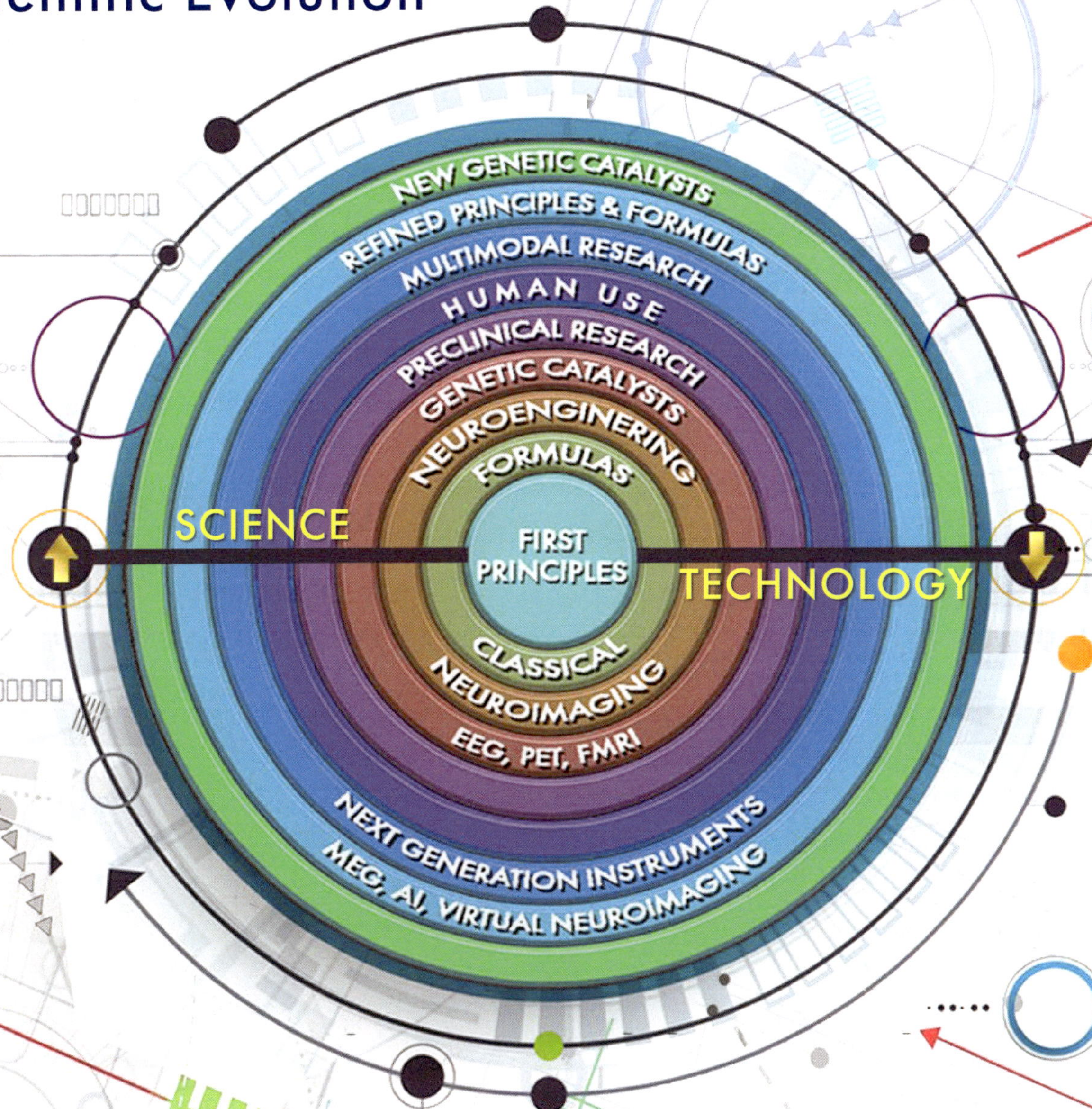

9.2 Ethical Use

The emergence of cognitive enhancement technologies is inevitable. These inventions will profoundly shape the future of human potential, and consequently, they carry a substantial responsibility.

To ensure these advancements serve highest good, a conscious and principled leadership group is being formed to provide ethical stewardship. This group will include visionaries, ethicists, scientists, clinicians and engineers dedicated to safe-guarding human dignity, agency, and well-being.

Their mission will be to establish a framework of guiding principles that transcend regulatory compliance and prioritize integrity, compassion, and long-term societal benefit. They will define ethical benchmarks for research, development, and application which consider safety, efficacy, accessibility, fairness, and the preservation of mental autonomy. Through a transparent and inclusive process, this leadership body will offer moral clarity in an age of accelerating technological change.

While we cannot prevent every misuse or unintended consequence, we can establish uniform standards of excellence. Providers, institutions, and technologies that uphold these standards can be recognized through a formal system of ethical certification. This recognition will empower users to make informed, value-aligned decisions while incentivizing responsible innovation across this new industry.

The goal is consciously guide the future to ensure that cognitive enhancement becomes a force for healing, growth, and the flourishing of human consciousness.

.3 Further Exploration

y treating consciousness as an lement independent of time and pace, metacognitive neurophysics rovides a framework for nderstanding the irrefutable vidence for psi phenomena which onventional neuroscience cannot xplain. In this sense, it is aligned vith a larger truth.

levertheless, any manuscript roposing both a new science and a ew stage of human evolution will ivite scrutiny. While the Sapients rchitecture rests on sound principles, everal areas deserve open discussion nd exploration:

1. Scientific Validity

Every great scientific transformation has begun with ideas that seemed speculative or even implausible at first. Metacognitive neurophysics extends neuroscience into new domains of consciousness. Critics may question whether constructs such as awareness packets or wave-field coupled circuits are empirically verifiable. We welcome this dialog. These hypotheses are offered as a scaffolding for experimental design, not as dogma. Future research must either validate or refine these models.

2. Genetic Enhancement

Early studies suggest remarkable gains in memory and reductions in anxiety, but replication and long-term studies are needed. Translating preclinical outcomes into safe, reproducible human therapies is a major frontier. These techniques should be viewed as promising prototypes rather than finished solutions.

3. Inclusivity and Social Concerns

Introducing genetic upgrades raises legitimate concerns of misuse or inequality. The Sapients ethos explicitly rejects any elitist or exclusionary application. Its principles are rooted in access, inclusivity, and the responsible use of biotechnology. It calls for robust institutional frameworks to govern distribution, safety, and fairness.

4. Scope of Impact

The manuscript envisions Sapients as catalysts for a new golden age. While aspirational, these outcomes cannot be guaranteed. They serve as guiding stars rather than fixed predictions. Real progress will depend on both scientific rigor and the ethical character of the individuals and communities who embrace these tools.

5. Conclusions

By acknowledging these critiques openly, we hope to invite dialogue, collaboration, and rigorous inquiry. *Sapients* should not be seen as a finished doctrine, but as a living blueprint — one that evolves alongside the evidence, technologies, and values of humanity.

9.4 New Awareness Science

Hundreds of experiments indicate our current science is incomplete; a subset of a larger understanding which embraces psi phenomena (e.g., telepathy, clairvoyance, precognition, ESP, OBE's, synchronicity). A new science is needed which encompasses both conventional materialism as well as the reality of psychic experiences.

Towards this end, Dr. Dean Radin's first book, *The Conscious Universe,* posits that consciousness has both local and non-local characteristics, and that non-local consciousness by definition has no location in space or time. Besides applying the ideas of quantum non-locality to consciousness, *Conscious Universe* also introduces a "bimodal" causation model of reality, which proposes that brain activity interacts with consciousness through neurology (bottom-up causation), and that consciousness can also influence brain activity (top-down causation).

This model provides an insightful entry point into understanding the mind/matter interface.

Radin's third book, *Supernormal,* presents evidence that meditation increases psi abilities. Psi abilities are characteristic of non-local consciousness. Among other effects, meditation is known to lower brainwave activity down into the lower-frequency alpha and theta states. Meditators in these states who demonstrated increased psi abilities provided evidence that increased non-local consciousness (which has no wavelength) is correlated with reduced brainwave activity (which has wavelength).

What would a new consciousness science that encompassed psi phenomena look like?

A new consciousness science that encompassed psi phenomena would have to satisfy three major requirements.

1

Consider consciousness as irreducible.

Consciousness is a fundamental element of the universe.

2

Distinguish pure awareness from consciousness.

Consciousness is information-based and content-driven whereas pure awareness is a presence without information content.

3

Codify the mind/matter interface.

As *Conscious Universe* implies, brainwaves are the interface zone between consciousness and physiology. Since brain-waves are governed by the laws of physics, explaining their interactions with consciousness will require mathematics.

There are over 200 theories of consciousness to choose from. Will any of them fit the bill?

Landscape of Consciousness

New Awareness Science

A LANDSCAPE OF CONSCIOUSNESS

Robert Lawrence Kuhn

1. MATERIALISM

1. Philosophical
01 eliminative
02 epiphenomenalism
03 functionalism
04 emergence
05 identity
06 Searle
07 Block
08 Flanagan
09 Papineau
10 Goldstein
11 Hardcastle
12 Stoljar

2. Neurobiological
01 Edelman
02 Crick-Koch
03 Baars
04 Dennett
05 Minsky
06 Graziano
07 Prinz
08 Sapolsky
09 Mitchell
10 Bach
11 circuits/cycles
12 Northoff
13 Bunge
14 Hirstein

3. Electromagnetic
1 Jones
2 Pockett
3 McFadden
4 ephaptic
5 Ambron
6 Llinas
7 Zhang

4. Computational & Informational
1 computational
2 Grossberg
3 complex/adaptive
4 critical brain
5 Pribram
6 Doyle
7 emergent info
8 mathematical

5. Homeostatic & Affective
01 predictive
02 Seth
03 Damasio
04 Friston
05 Solm
06 Carhart-Harris
07 Buzsáki
08 Deacon
09 Pereira
10 Mansell
11 projective
12 Pepperell

6. Embodied & Enactive
1 embodied
2 enactivism
3 Varela
4 Thompson
5 blind spot
6 Bitbol
7 direct
8 Gibson

7. Relational
1 A.Clark
2 Noë
3 Loorits
4 Lahav
5 Tsuchiya
6 Jaworski
7 process

8. Representational
01 first-order
02 Lamme
03 higher-order
04 Lau
05 LeDoux
06 Humphrey
07 Metzinger
08 Jackson
09 Lycan
10 trasnparency
11 Tye
12 Thagard
13 T.Clark
14 Deacon

9. Language
1 Chomsky
2 Searle
3 Koch
4 Smith
5 Jaynes
6 Parrington

10. Phylogenetic
1 Dennett
2 LeDoux
3 Jablonka
4 value
5 Andrews
6 Reber
7 Feinberg/Mallatt
8 Levin
9 James

2. NON-REDUCTIVE PHYSICALISM

1 Ellis
2 Murphy
3 van Inwagen
4 Nagasawa
5 Sanfey
6 Northoff

3. QUANTUM

01 Penrose-Hameroff
02 Stapp
03 Bohm
04 Pylkkänen
05 Wolfram
06 Beck-Eccles
07 Kauffman
08 Torday
09 Smolin
10 Carr
11 Faggin
12 Fisher
13 Globus
14 Poznanski
15 extensions
16 Rovelli

4. INTEGRATED INFORMATION THEORY

1 critiques
2 Koch

5. PANPSYCHISMS

01 micropsychism
02 panprotopsychism
03 cosmopsychism
04 qualia force
05 qualia space
06 Chalmers
07 Strawson
08 Goff
09 A.Harris
10 Sheldrake
11 physics
12 Whitehead

6. MONISMS

01 Russellian
02 Davidson
03 Velmans
04 Strawson
05 Polkinghorne
06 Teilhard
07 Atmanspacher
08 Ramachandran
09 Tegmark
10 QRI valence
11 Bentley Hart
12 Leslie

7. DUALISMS

01 property
02 traditional
03 Swinburne
04 composite
05 Stump
06 Feser
07 Moreland
08 interactive
09 emergent
10 Kind
11 Jewish
12 Christian
13 Islamic
14 god
15 Indian
16 Indigenous
17 soul realms
18 Theosophy
19 Steiner
20 nonphysical

8. IDEALISMS

01 Indian
02 Buddhism
03 Dao De Jing
04 Kastrup
05 Hoffman
06 McGilchrist
07 Chopra
08 universe
09 Goswami
10 Spira
11 Nader
12 Ward
13 Albahari
14 Meijer
15 imaginative

9. ANOMALOUS & ALTERED STATES

01 Bergson
02 Jung
03 Radin
04 Tart
05 Josephson
06 Wilber
07 Combs
08 Schooler
09 Sheldrake
10 Grinberg
11 Graboi
12 NDEs/survival
13 DOPS
14 Bitbol
15 Campbell
16 Hiller
17 Harp
18 Swimme
19 Langan
20 meditation
21 psychedelic

10. CHALLENGE

01 Nagel
02 McGinn
03 S.Harris
04 Eagleman
05 Tallis
06 Nagasawa
07 Musser
08 Davies

Figure created by Alex Gomez-Marin

Landscape of Consciousness [1]

New Awareness Science

Seven of the 209 theories in Robert Kuhn's Landscape of Consciousness contain mathematical treatments of consciousness:

1. Integrated Information Theory (IIT) - Uses Φ (phi) to measure consciousness.
2. Friston's Free-Energy Principle - Models consciousness using Bayesian inference and entropy minimization.
3. Hoffman's Conscious Realism - Represents consciousness as a Markovian interaction network.
4. Mathematical Ontology of Consciousness - Uses Hilbert space tensor models to define conscious states.
5. Duan's Mathematical "Bug" Theory - Describes consciousness as information oscillations modeled using Fourier transforms.
6. Projective Consciousness Model (PCM) - Applies projective geometry to describe conscious perception.
7. Wolfram's Ruliad Approach - Defines consciousness as a hypergraph of computational relationships.

Three of these seven theories consider consciousness as a fundamental element of the universe independent of matter:

1. Hoffman's Conscious Realism (Idealism) - Consciousness is primary, and physical reality is an emergent phenomenon.
2. Mathematical Ontology of Consciousness (Platonism / Monism / Panpsychism) - Consciousness exists within mathematical structures.
3. Wolfram's Ruliad Approach (Computational Panpsychism / Mathematical Monism) - Consciousness is an inherent feature of the computational universe.

[1] A landscape of consciousness: Toward a taxonomy of explanations and implications
Progress in Biophysics and Molecular Biology 190 (2024) 28–169
Robert Lawrence Kuhn, January, 2024

Landscape of Consciousness

New Awareness Science

All three mathematically-based theories that consider consciousness as fundamental treat it as information based and content-driven.

However, pure awareness is distinct from consciousness.
- Consciousness is information-based and content-driven.
- Awareness, in its purest form, is contentless presence.

None of these three theories treat contentless pure awareness.

Advaita Vedanta is the only framework that explicitly distinguishes pure awareness from content-based consciousness, but it does not contain mathematics.

Conclusion:
- Existing mathematical models of consciousness rely on structured information processing.
- No scientific or mathematical theory in Kuhn's landscape explicitly models formless, contentless awareness.

Requirements for a Science of Awareness
1. Consider consciousness as irreducible.
2. Treat pure awareness as contentless.
3. Explain how consciousness and brainwaves interact mathematically.

Adding a new theory which meets these requirements to Kuhn's landscape yields a chart like this:

Landscape of Consciousness

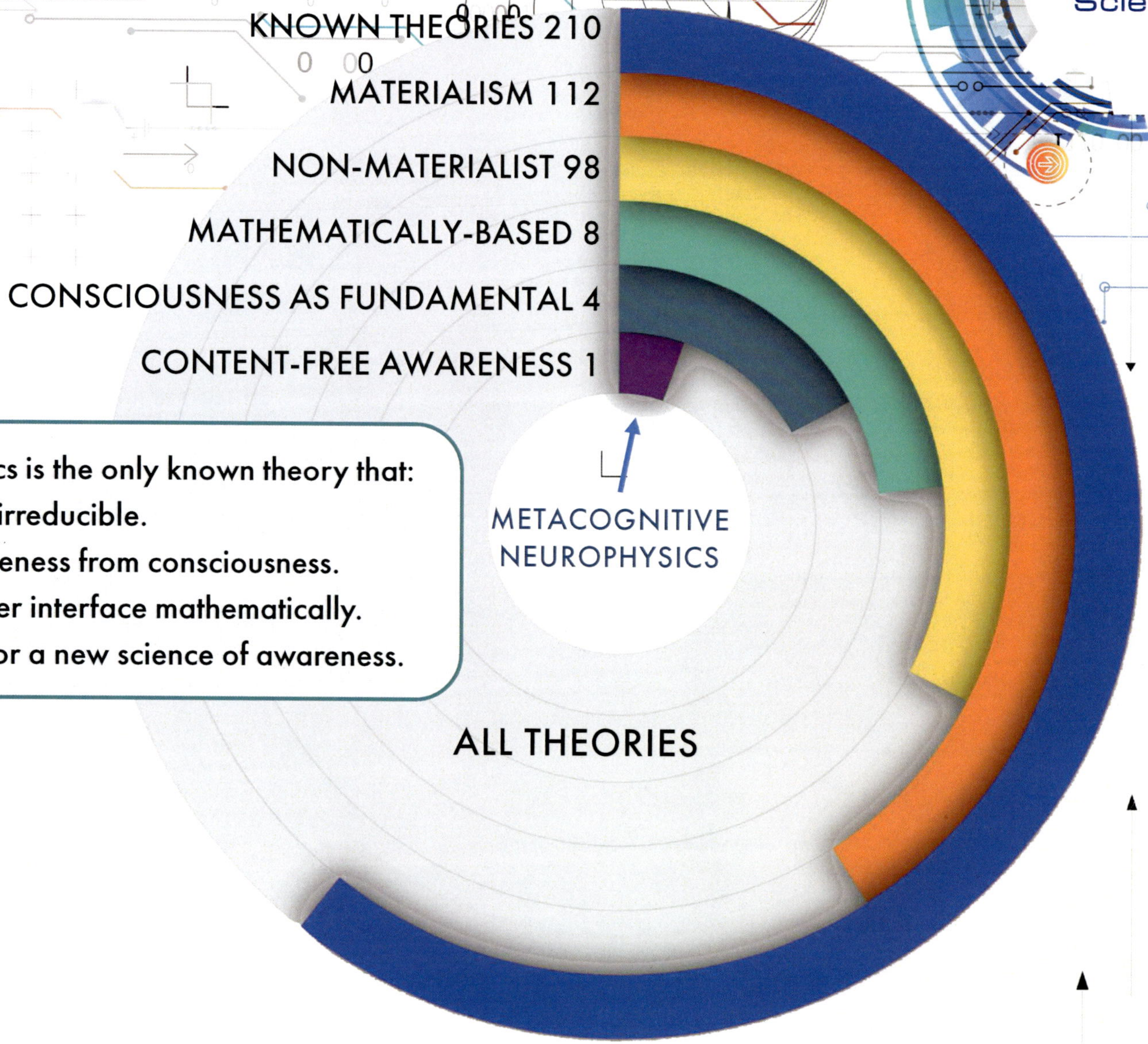

Metacognitive neurophysics is the only known theory that:

1. Treats consciousness as irreducible.
2. Distinguishes pure awareness from consciousness.
3. Codifies the mind/matter interface mathematically.

It meets the requirements for a new science of awareness.

Memory, Consciousness and Electromagnetic Field Dynamics: A Unified Framework

Unified Framework Theory

Overview

Metacognitive neurophysics provides a unified theoretical framework integrating memory theory, state-dependent recall, and consciousness through the lens of electromagnetic (EM) field dynamics.

Building on neuroscience, field theory, and non-emergent consciousness assumptions, this model asserts that memories are inert until partially activated, that consciousness has no wave signature, and that reconsolidation occurs through wave-field modulation.

This framework reconciles the gap between brain-based activation and the fieldless, nonlocal nature of conscious awareness.

Introduction

The relationship between memory, consciousness, and brain activity is traditionally approached through neuroanatomy and emergent materialism. An alternative view is proposed where:

- Consciousness is a fundamental, non-emergent aspect of reality.
- Memories are stored in latent, non-active form and generate electromagnetic signatures only upon activation.
- Conscious awareness is distinct from EM field activity and represents a fieldless, formless witnessing substrate.

Memory Field Signatures and Resonance

Unified Framework Theory

Memory and EM Field Signatures

1

Memories are encoded via neural activity, forming distinct EM field configurations. These field patterns correspond to distributed sensory, emotional, and procedural traces.

- Dormant Memory: No EM field. Exists as a latent structural configuration.
- Partial Activation: Cue-driven local field resonance emerges. May remain subconscious.
- Full Activation: Field coherence allows global resonance. May or may not enter consciousness.

State-Dependent Memory as Field Resonance

2

State-dependent recall occurs when the brain's global EM field aligns with the latent field pattern of a stored memory. This resonance model explains:

- Why specific moods or brain states improve recall (state match = field match).
- How subconscious influences arise from partial field overlaps.
- That consciousness is not required for activation, only for awareness.

Consciousness and Memory

Unified Framework Theory

Consciousness as Fieldless Awareness

3

This model assumes that conscious awareness:

- Has no wavelength, frequency, or EM geometry.
- Is non-local, timeless, and unconditioned.
- Collapses field activity upon entry—memories become formless when witnessed.

Thus, any memory that is emitting a field is, by definition, not presently in pure consciousness. Consciousness is a substrate that allows memory to become experientially present but does not itself contain wave-based content.

Memory Reconsolidation and EM Field Collapse

4

According to memory reconsolidation theory:

- Conscious awareness destabilizes the original memory field.
- New input (emotional, cognitive, therapeutic) modifies the field structure.
- Upon exit from consciousness, a new EM field pattern is written—constituting reconsolidation.

This model reframes reconsolidation as a field metamorphosis mediated by formless awareness.

Layered Model: Subconscious Memory Influencing State-Dependent Accessibility

Unified Framework Theory

Layers 1 and 2

5

This layered model illustrates how subconscious memory, through EM field modulation, affects memory accessibility:

Δ Layer 1: Dormant Memory Storage

- Stored as synaptic configurations and network topology.
- No electromagnetic field activity while inactive.

Δ Layer 2: Subconscious Partial Activation

- Triggered by contextual, sensory, or emotional cues.
- Generates a localized EM field.
- Influences mood, perception, and behavior subtly without entering awareness.

Layers 3 and 4

5

Δ Layer 3: Global Brain State and Field Configuration

- Partial activations accumulate to alter the global EM field.
- A new field configuration forms based on the brain's oscillatory state (e.g., theta, alpha, beta).
- This altered field state may match other latent memory traces, enabling recall.

Δ Layer 4: Resonance and Recall

If global EM field geometry aligns with a latent memory's encoding field, resonance occurs. Memory becomes accessible either subconsciously or consciously, depending on activation strength and relevance.

Conclusions

Unified Framework Theory

Feedback Loop

5

Δ Feedback Loop
Subconscious activations → field state modulation → altered recall profile → reinforcement or reconsolidation → new field dynamics.

This model underscores that memory retrieval is not linear or fixed—it emerges from dynamic electromagnetic coherence shaped by subconscious memory fragments.

Implications
- Procedural memory is activated and influences behavior without entering conscious awareness.
- Subconscious memory corresponds to partially active field patterns that shape mood, perception, and bias.

Conclusions

6

- Conscious recall involves field collapse, integration, and rewriting.
- Psychotherapy works by exploiting this collapse-and-reform cycle through emotional and cognitive interventions held in awareness.

This framework suggests a layered ontology:
- Consciousness as a non-oscillatory substrate
- EM fields as the medium of mental form and influence
- Memory as a latent potential that becomes field-active when contextually resonant

By integrating neuroscience, field theory, and non-emergent consciousness, this model offers a novel map for understanding memory, awareness, and transformation.

Chapter 10: Reference

- Experimental Results
- Interactive Simulation Model
- Neural Electrodynamics
- Experimental Results Interpretation
- Current Density
- Neuron Signaling
- Brain Frequencies Cognitive Traits
- Brainwave Frequency

Experimental Results

Cognitive neuroengineering has been validated in multiple indications (1) in multiple species (2) using multiple gene editors (3) on multiple targets (4) for multiple genes (5).

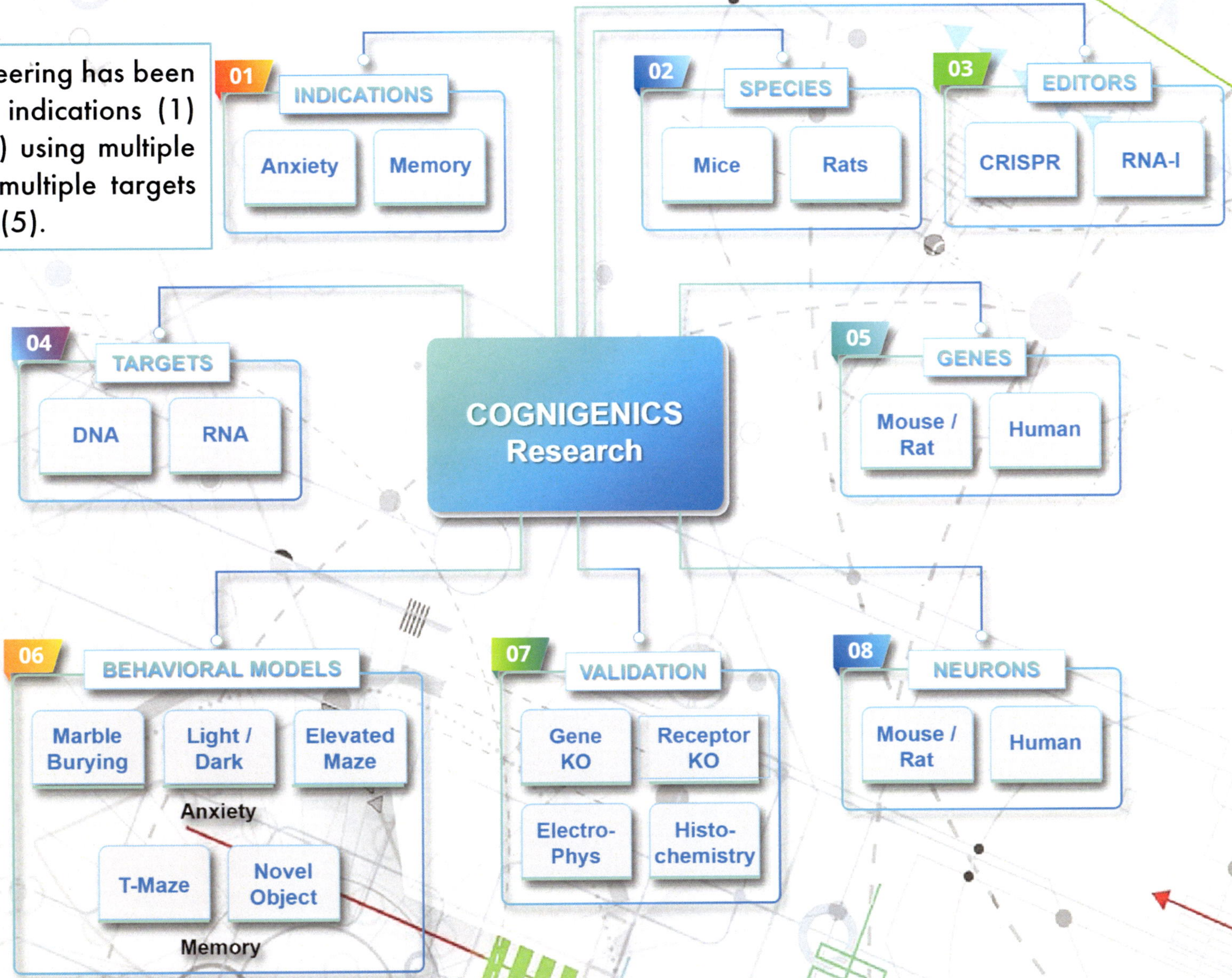

Results have been replicated across multiple behavioral models (6) and validated in multiple diagnostic assays (7) on multiple neuron types (8).

Metacognitive Neurophysics Interactive Model

COGNITIVECOLLEGE.ORG – SCIENTIFIC TAB

Model Navigation

1. Firing Rates | 2. Neural Currents | 3. Impedance Analysis | 4. Brainwave Amplitude

Brainwave Frequencies & Neuron Firing Rates

Average individual neuron firing rates across 10 frequency bands (Delta through Beta)

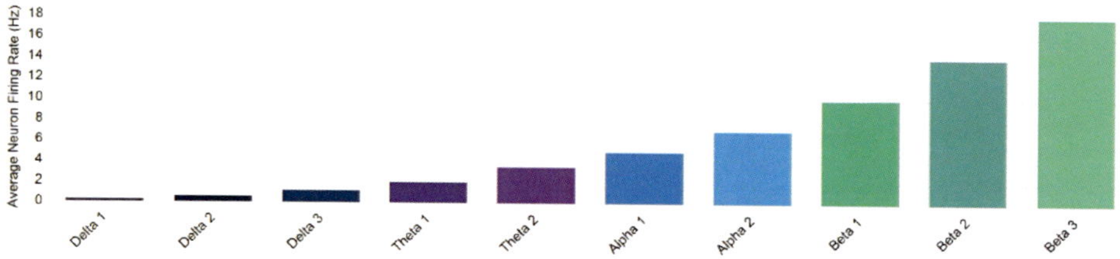

Model Navigation

1. Firing Rates | 2. Neural Currents | 3. Impedance Analysis | 4. Brainwave Amplitude

Neuronal Firing Rates & Driven Currents

RC Circuit Model: Current calculations from individual neuron firing rates

5-HT2A Receptor Density: 100%

Baseline = 100%. Reduced by genetic downregulation

R_synaptic: 100 MΩ

Membrane Resistance (R_membrane): 50 MΩ

Range: 10-100 MΩ (typical neurons, CONSTANT)

Hide Advanced Metrics

Circuit Parameters: ΔV = 110 mV/spike, Spike Duration = 1.5 ms, R_synaptic = 100 MΩ, R_membrane = 50 MΩ, R_total = 150 MΩ

Peak Current per Spike

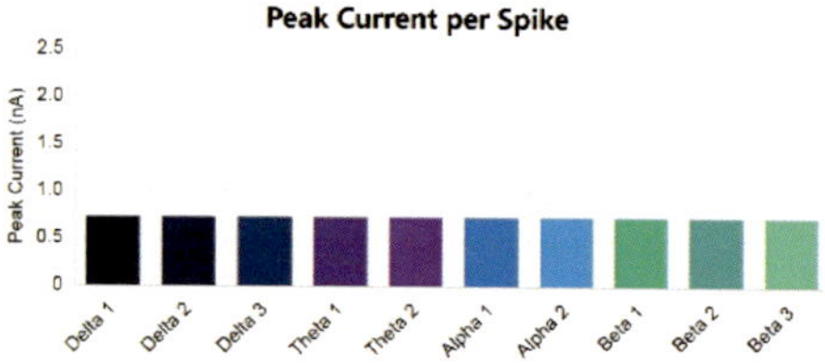

I_peak = ΔV / R_total = 0.733 nA

Time-Averaged Current

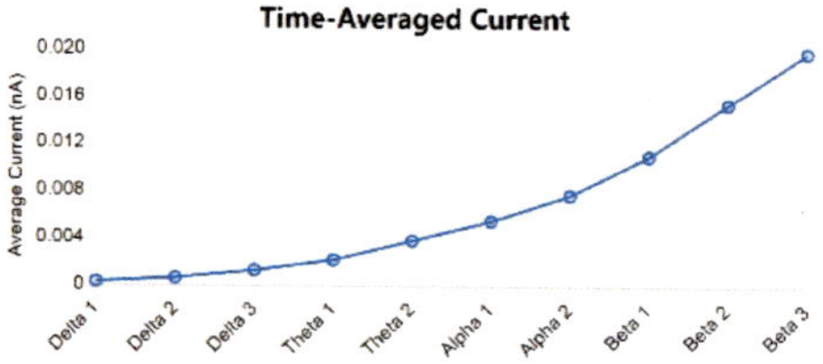

I_avg = I_peak × (spike duration × firing rate)

Neuronal Impedance & Capacitive Reactance Analysis

Two-resistance RC circuit model: Z_total = R_synaptic + Z_membrane, where X_C is calculated at brainwave frequency (1-30 Hz)

Impedance Components vs Frequency

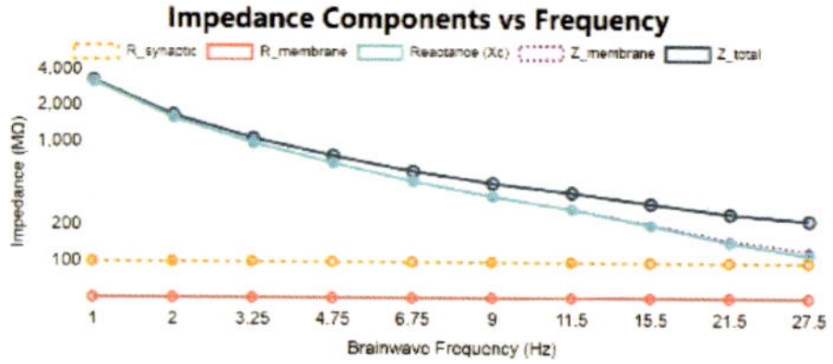

Log scale: X_C = 1/(2πfC), Z_membrane = √(R_membrane² + X_C²), Z_total = R_synaptic + Z_membrane

Peak Current: Resistive vs Impedance

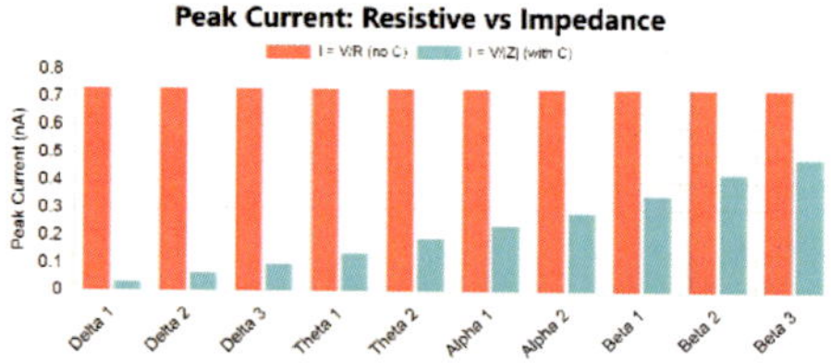

Capacitance reduces current at low frequencies

5-HT2A Receptor Density: 100%

Baseline = 100%. Reduced by genetic downregulation

R_synaptic: 100 MΩ

Membrane Resistance (R_membrane): 50 MΩ

Typical range: 10-100 MΩ (CONSTANT)

Membrane Capacitance: 0.05 nF

Typical range: 0.01-0.2 nF (10-200 pF, CONSTANT)

RC Time Constant (τ): 2.50 ms
Cutoff Frequency (f_c): 63.66 Hz

1. Firing Rates | 2. Neural Currents | 3. Impedance Analysis | 4. Brainwave Amplitude

5-HT2A Receptor Density: 100%

Baseline = 100%. Reduced by genetic downregulation

R_synaptic: 100 MΩ

Membrane Resistance (R_membrane): 50 MΩ

Typical range: 10-100 MΩ (CONSTANT)

Membrane Capacitance: 0.05 nF

Typical range: 0.01-0.2 nF (10-200 pF, CONSTANT)

RC Time Constant (τ): 2.5(
Cutoff Frequency (f_c): 63

Phase Angle (Current vs Voltage)

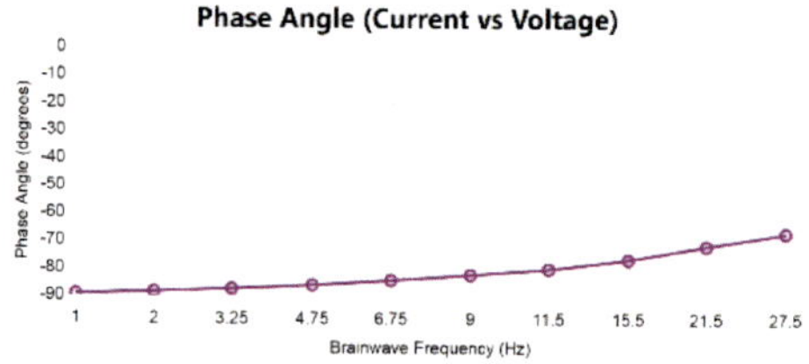

φ = arctan(-X_C/R): Current lags voltage in capacitive circuits

Time-Averaged Current (with Firing Rate)

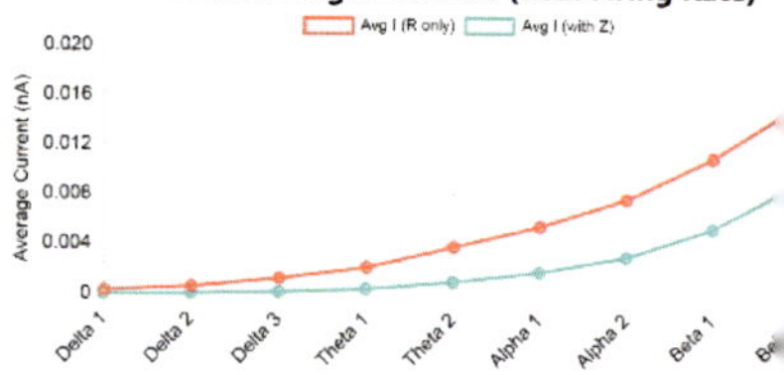

I_avg = I_peak × (spike duration × firing rate)

Model Navigation

1. Firing Rates | 2. Neural Currents | 3. Impedance Analysis | 4. Brainwave Amplitude

Brainwave Amplitude from Neural Current Density

Calculating macroscopic EEG amplitudes from microscopic neural currents on pathways with edited neurons.

Local Brainwave Amplitude

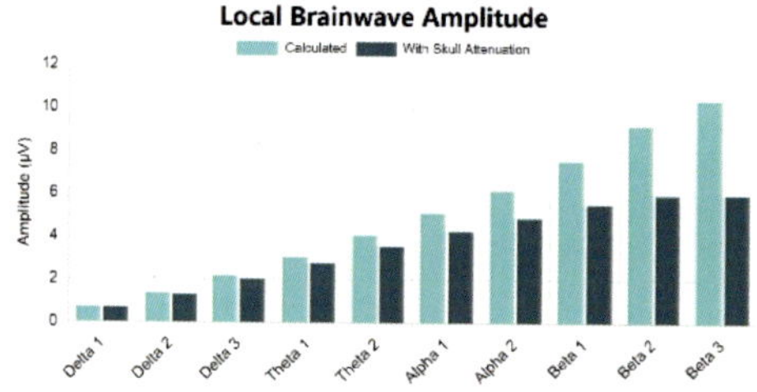

V = (I × ρ) / (4π r²), includes synchrony and orientation factors

Amplitude Spectrum

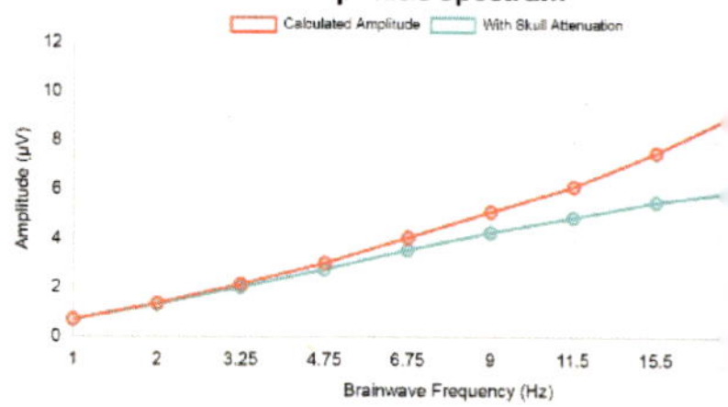

Peak-to-peak amplitude vs brainwave frequency

5-HT2A Receptor Density: 100%

Baseline = 100%. Reduced by genetic downregulation

R_synaptic: 100 MΩ

Membrane Resistance (R_membrane): 50 MΩ

Membrane Capacitance: 0.05 nF

Active Neurons: 1.0M

Synchrony: 40%

Fraction of neurons fir

Source-to-Scalp Distance: 2.5 cm

Total Synchronized Current

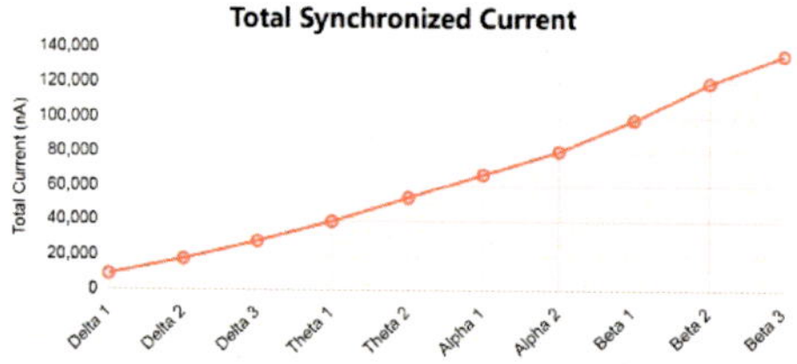

Sum of synchronized neuron currents across bands

Power Spectral Density

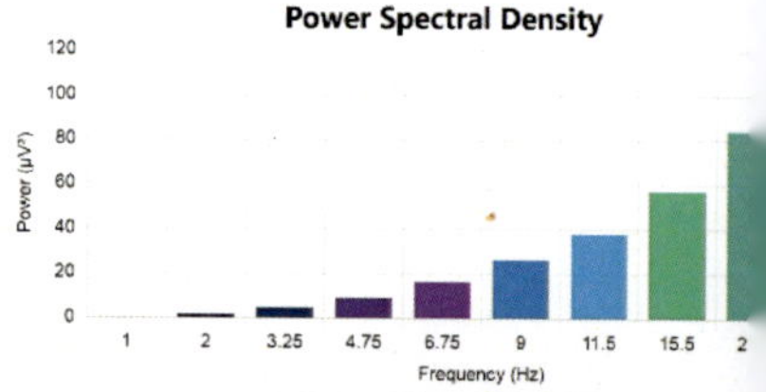

Power = Amplitude² (proportional to energy)

Neural Electrodynamics

How Receptor Density Controls Brain Current and Awareness

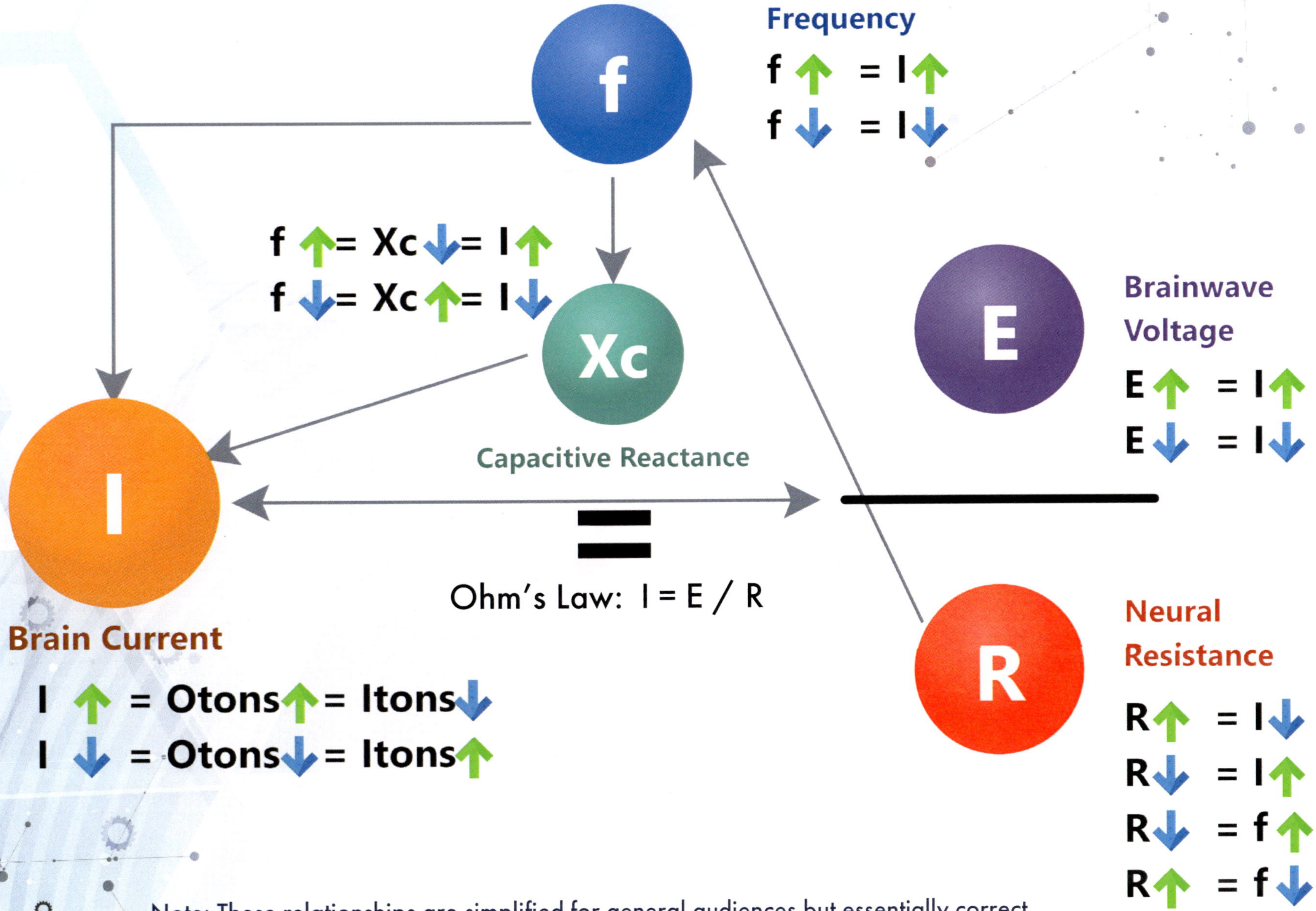

Note: These relationships are simplified for general audiences but essentially correct.

↑ = Increase ↓ = Decrease

Experimental Results Interpretation

Current vs. Impedance for 20W

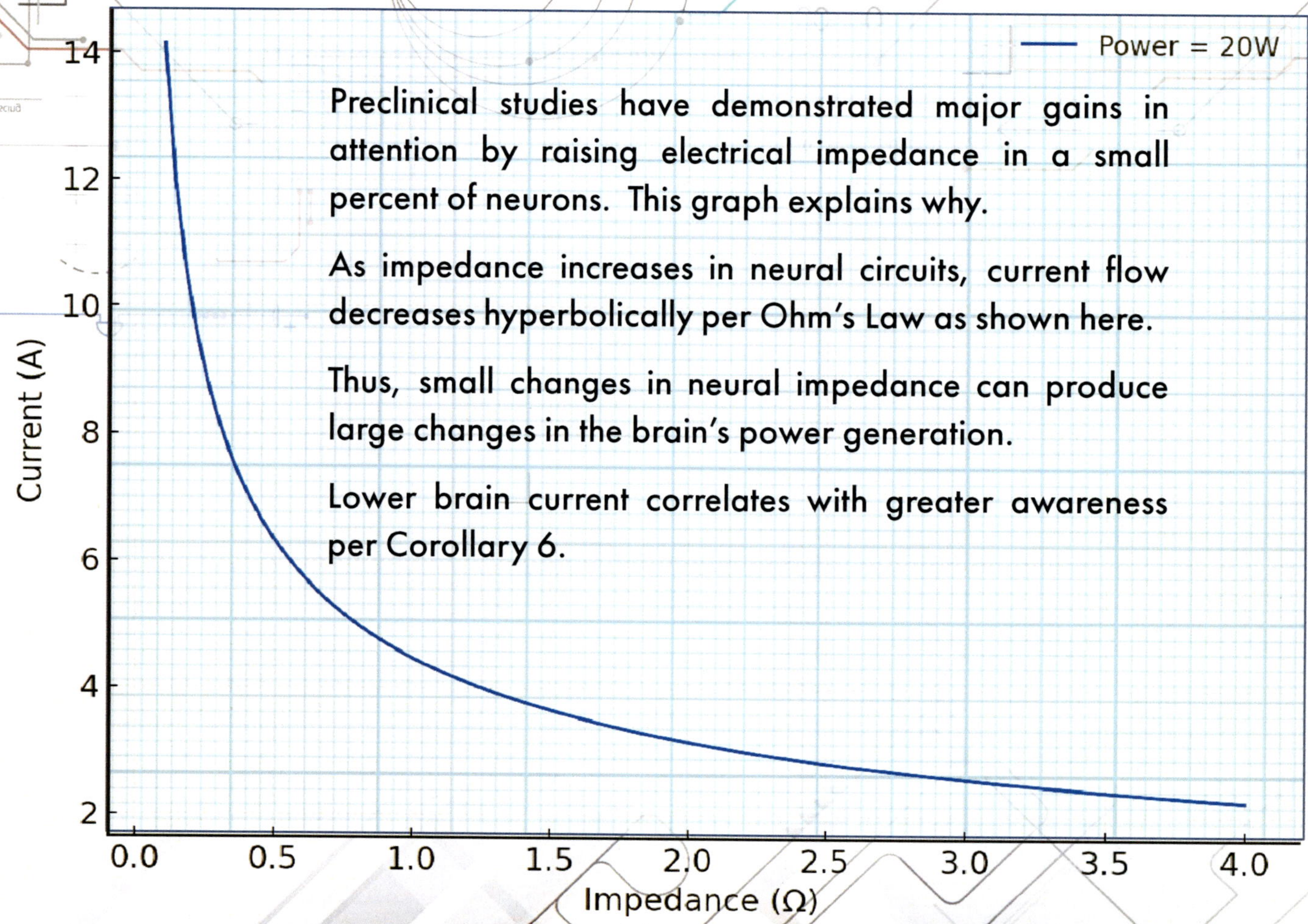

Preclinical studies have demonstrated major gains in attention by raising electrical impedance in a small percent of neurons. This graph explains why.

As impedance increases in neural circuits, current flow decreases hyperbolically per Ohm's Law as shown here.

Thus, small changes in neural impedance can produce large changes in the brain's power generation.

Lower brain current correlates with greater awareness per Corollary 6.

Ohm's Law

Current Density

Limbic vs. Cortical Regions

1 The limbic system generates denser, more concentrated currents per unit volume during memory encoding, while cortical networks produce weaker but more widespread signals

2 The chart compares representative local field potential (LFP) amplitudes measured in limbic versus cortical regions during memory processes.

3 In hippocampus and amygdala (limbic structures), memory encoding produces high-amplitude sharp-wave ripple events (SWRs) in the range of ~150–200 μV, reflecting highly synchronized bursts of synaptic current.

4 In contrast, corresponding ripple-like activity in cortical networks typically falls in the ~50–80 μV range.

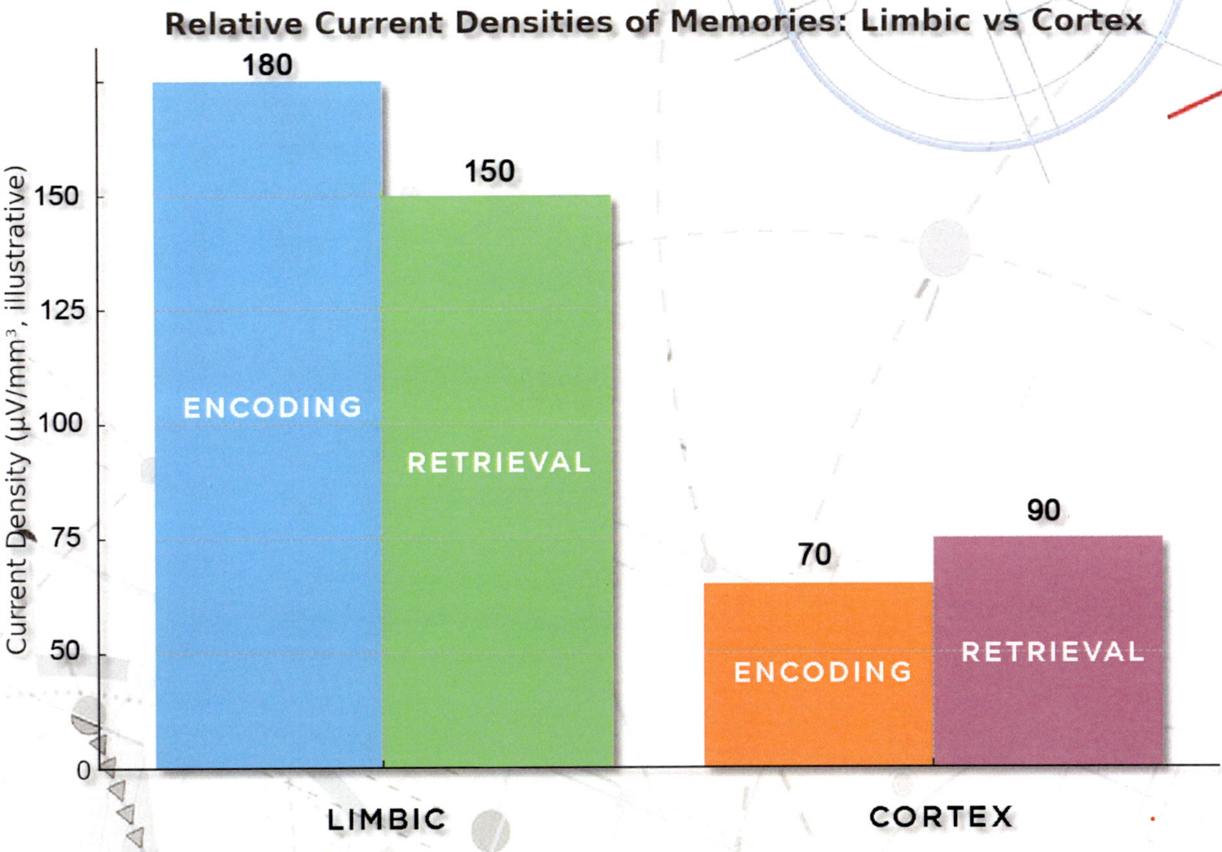

Encoding: Limbic amplitudes peak (~150–200 μV)
Retrieval: Limbic remains strong (~120–180 μV),

Encoding: Cortex lower (~50–80 μV).
Retrieval: Cortex rises slightly (~70–100 μV), reflecting its role in detailed recall.

Current Densities of Memories References

Buzsáki, G. (2015). Hippocampal sharp wave–ripple: A cognitive biomarker for episodic memory and planning. Hippocampus, 25(10), 1073–1188. https://doi.org/10.1002/hipo.22488

Dickey, C. W., Sargsyan, A., Madsen, J. R., Eskandar, E. N., Cash, S. S., Halgren, E., & Mehta, A. D. (2022). Cortical ripples during NREM sleep and waking in humans. Journal of Neuroscience, 42(42), 7931–7946. https://doi.org/10.1523/JNEUROSCI.0609-22.2022

Dickey, C. W., Sargsyan, A., Madsen, J. R., Eskandar, E. N., Cash, S. S., Halgren, E., & Mehta, A. D. (2022). Widespread ripples synchronize human cortical activity. Proceedings of the National Academy of Sciences, 119(27), e2202537119. https://doi.org/10.1073/pnas.2202537119

Dickey, C. W., Eskandar, E. N., Madsen, J. R., Halgren, E., Mehta, A. D., & Cash, S. S. (2024). Human ripples link hippocampus, cortex, and thalamus at rest and in memory. PLOS Biology, 22(2), e3002524. https://doi.org/10.1371/journal.pbio.3002524

Norman, Y., Yeagle, E. M., Khuvis, S., Harel, M., Mehta, A. D., & Malach, R. (2021). Hippocampal ripples and their coordination with neocortical activity during human memory retrieval. Neuron, 109(2), 402–413.e5. https://doi.org/10.1016/j.neuron.2020.10.032

Pedrosa, R. P., Clawson, B. C., Bakhurin, K. I., Latchoumane, C.-F. V., & Halgren, E. (2022). Coordination of hippocampal sharp wave ripples and neocortical activity during human memory consolidation. Proceedings of the National Academy of Sciences, 119(10), e2117565119. https://doi.org/10.1073/pnas.2117565119

Sakon, J. J., Munoz, C., Cash, S. S., Halgren, E., & Mehta, A. D. (2024). Hippocampal ripples track episodic memory encoding and retrieval in humans. Journal of Neuroscience, 44(6), 1031–1045. https://doi.org/10.1523/JNEUROSCI.1822-23.2023

Costa, K. M., Giocomo, L. M., Staresina, B. P., & Hauk, O. (2022). Aversive memory formation in humans involves an amygdala–hippocampus network. Nature, 610, 345–351. https://doi.org/10.1038/s41586-022-05228-0

Qasim, S. E., Fried, I., & Jacobs, J. (2021). Neuronal activity in the human amygdala and hippocampus enhances emotional memory encoding. bioRxiv. https://doi.org/10.1101/2021.02.17.431672

Butt, M., Almourad, M. B., & Hamid, Q. (2019). Brain signals of posttraumatic stress disorder filtered using EEG and MEG: A review. Frontiers in Human Neuroscience, 13, 133. https://doi.org/10.3389/fnhum.2019.00133

Radiske, A., Gonzalez, M. C., Rossato, J. I., Köhler, C. A., Bevilaqua, L. R., & Cammarota, M. (2020). Cross-frequency phase–amplitude coupling between theta and gamma rhythms in the amygdala–hippocampus is enhanced during avoidance memory reconsolidation. eNeuro, 7(1), ENEURO.0465-19.2020. https://doi.org/10.1523/ENEURO.0465-19.2020

Neuron Signaling
Axiom 7

Only 1-12% of the brain's neurons are normally active at a time.

This active neuron population scales with frequency. At an average brainwave frequency of 30 Hertz, four times as many neurons are likely to be signaling at any given instant than at an average rate of 4 Hertz.

Refer to the next page for details.

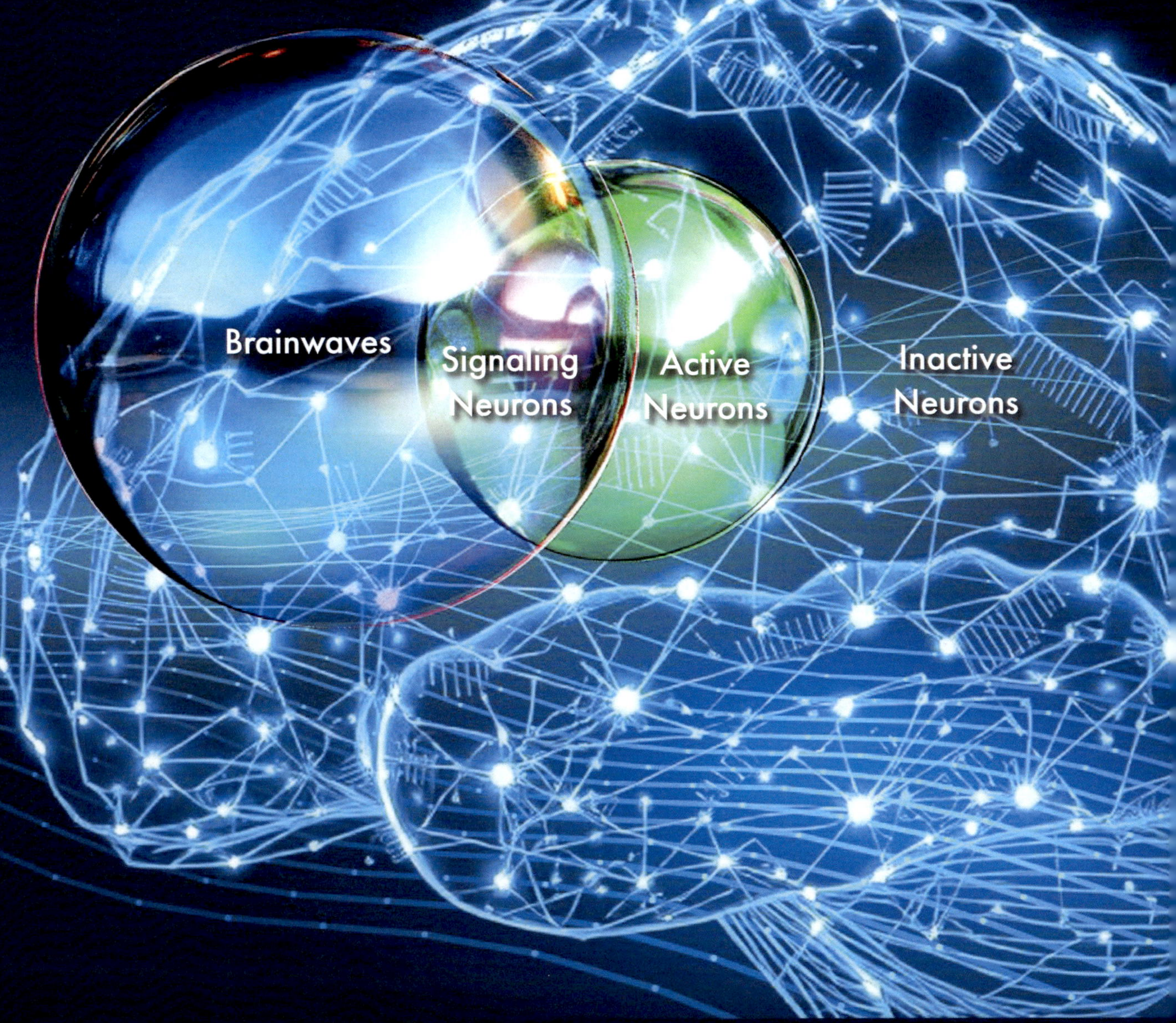

Neuron Signaling Analysis

EEG Band	Frequency Range (Hz)	Temporal Resolution (ms)	% of Neurons Active per second	% of Neurons Signaling (instantaneous)	% of Signaling Neurons that Synchronize	Net % of Total Neurons Driving the Wave
Delta	1–4	250–2000	1–3%	0.5–1%	40–50%	0.5–1%
Theta	4–8	125–250	3–6%	1–3%	40–60%	1–3%
Alpha	8–12	83–125	4–8%	2–4%	40–60%	2–4%
Beta	13–30	33–77	8–12%	4–7%	50–60%	4–7%

- Temporal resolution: How often excitability windows open each second (faster rhythms = finer timing).
- Active neurons: Percent that spike at least once per second (cycle-dependent).
- Signaling neurons: Percent spiking at the exact instant of measurement.
- Synchronization: The fraction of signaling neurons that align in phase.
- Net driving neurons: The effective percent of total neurons that contribute coherently to the oscillation.

Brain Current/Frequency Cognitive Traits

This table plots cognitive traits by brain current and brainwave frequency. It suggests that higher awareness and metacognitive states tend to correlate with lower brainwave frequencies and power levels. More encapsulated awareness states are present at higher power and frequency levels.

Hard science data supports most, but not all, of the table's underlying assumptions, so it must be regarded as theoretical. However, its overall structure seems to recapitulate the Law of Conservation of Awareness and reinforce Corollaries 4 and 6. Descriptions of each state appear on the next page.

	Current →	10–11 W	11–15 W	16–20 W	21–25 W
	Frequency ↓	1	2	3	4
A	Delta (1–4 Hz)	Grounded Stillness Deep meditative awareness	Somatic Calm Awareness of physical sensations	Dreamlike Drift Associative thinking, intuition	Fragmented Overload Scattered awareness, impulsiveness
B	Theta (4–8 Hz)	Reflective Access Curiosity, non-judging awareness	Insight Looping Imagination, emotional insight, mind-wandering	Emotional Immersion Appreciation, tenderness, empathy, reminiscence	Emotional Flooding Emotional overwhelm, self-referential looping
C	Alpha (8–13 Hz)	Present Focus Relaxed awareness, alert concentration	Flow Readiness Effortless action, stable concentration	Controlled Analysis Engaged analytical work	Cognitive Tension Over-monitoring hypervigilance
D	Beta (13–30 Hz)	Passive Readiness Calm vigilance, non-demanding awareness	Structured Focus Mental clarity, productive reasoning	Active Control Complex cognition, multi-tasking	Overactivation Mental noise, rigidity, impulsiveness

LAW OF CONSERVATION

ITONS OTONS

Brain Current/Frequency Cognitive Trait Descriptions

A1 – Grounded Stillness (10–11 W): This state reflects low-frequency synchronization with minimal current, typically observed in deep meditative absorption **or** slow-wave sleep.
A2 – Somatic Calm (11–15 W): As energy increases slightly, delta rhythms begin coupling with anterior cingulate and insular cortex in interoceptive states. Mindfulness practices and yoga produce this profile,
A3 – Dreamlike Drift (16–20 W): Mid-current delta states are likely engaging in REM-like or hypnagogic nonlinear memory retrieval, symbolic processing, or intuitive problem-solving.
A4 – Fragmented Overload (21–25 W): Delta synchrony with high current suggests overactive subcortical compensation without cortical regulation.

B1 – Reflective Access (10–11 W): Maps to access concentration, early-stage mindfulness, relaxed introspection and hippocampal coherence. Traits include curiosity, non-judging awareness, and receptive reflection.
B2 – Insight Looping (11–15 W): Theta becomes prominent in default mode and medial prefrontal activation. Traits: imagination, emotional insight, mind-wandering.
B3 – Emotional Immersion (16–20 W): Theta becomes affectively loaded, engaging limbic circuits. A warm, feeling-rich state where memories gain color and meaning. Traits: Appreciation, tenderness, empathy..
B4 – Emotional Flooding (21–25 W): High power theta links to maladaptive rumination or PTSD flashbacks. Activity in the PCC, vmPFC and parahippocampal gyrus spikes. Traits: emotional overwhelm, self-referential looping

C1 – Present Focus (10–11 W): Classic 'relaxed but alert' alpha state. Occipital and parietal alpha suppress distracting input. Seen in zen-style open awareness and focused concentration.
C2 – Flow Readiness (11–15 W): Alpha current supports efficient inhibition of non-relevant stimuli. allowing flow states to emerge. Traits: effortless action, graceful performance, stable concentration.
C3 – Controlled Analysis (16–20 W): Alpha begins coupling with frontal beta supporting task-switching, working memory, and light reasoning. Found in calm but engaged analytical work.
C4 – Cognitive Tension (21–25 W): High alpha power with high current suggests over-control or perfectionism... Traits include over-monitoring, hypervigilance, and effortful focus

D1 – Passive Readiness (10–11 W): Low beta at low power supports basic alertness. Traits: *calm vigilance*, *preparation*, *non-demanding awareness*.
D2 – Structured Focus (11–15 W): Moderate beta with moderate energy produces goal-directed, effective performance. Frontal lobe–PFC coordination. Traits: mental clarity, productive thinking, rationality.
D3 – Active Control (16–20 W): High beta supports complex cognition, planning, multi-tasking... High PFC and parietal involvement. Traits: deliberate reasoning, executive control, self-discipline.
D4 – Overactivation (21–25 W): Excess beta with peak current reflects stress, anxiety, and overthinking... Amygdala-PFC coupling becomes unbalanced. Traits: mental noise, rigidity, impulsiveness.

Brain Current/Frequency Cognitive Traits Analysis

1. Empirically Anchored Baseline and Extremes

• The estimates for deep meditation (~10W), resting state (~17W), and analytical mode (22–24W) are backed by fMRI, PET, and EEG findings, yielding an empirical physiological foundation for modeling brain current across awareness states.

• These endpoints make it possible to establish a nonlinear function mapping awareness levels to current or power generation, which underlies the frequency-current-state relationship seen in the table.

2. Conceptual Integration Across Domains

• The model integrates multiple domains into a holistic systems architecture, unifying disparate observations into a common conceptual framework.

o Neurophysiology (brainwaves, energy use),

o Cognitive neuroscience (states like flow, meta-awareness),

o Personality psychology (traits like conscientiousness or openness),

o Systems modeling (e.g., inverse frequency-power relationships).

3. Well-Justified Trait Placement

• Each cognitive trait (e.g., introspection, vigilance, mindfulness) is linked with a specific combination of brainwave frequency and current based on evidence from cognitive psychology, meditation studies, and EEG correlations, for example:

o High theta + low current → introspective, mindful states,

o High beta + high current → focused, vigilant traits,

o Low frequency + low current → transcendence, higher consciousness

• These placements are psychologically plausible and neurobiologically defensible.

Brainwave Frequency

Low-Frequency Brainwaves and Mental Clarity

Low-frequency brainwaves—specifically delta, theta, and alpha—are correlated with states of deep relaxation, inner stillness, and enhanced creativity. These patterns commonly emerge during meditation, deep sleep, and other altered states of consciousness. Their presence is associated with reduced anxiety, heightened clarity, and improved memory performance.

DMN Activity and Meditation

Functional MRI studies reveal that experienced meditators show reduced activity in the brain's default mode network (DMN), a network involved in mind-wandering and self-referential thought (Brewer et al., 2011). Suppressed DMN activity is closely linked to enhanced mindfulness and metacognitive awareness, contributing to improved emotional regulation and lower stress levels (Garrison et al., 2015; Pagnoni, 2012).

Altered States and Neural Activity

Altered states of consciousness—whether induced by meditation, psychedelics, or breathwork—often lead to a reduction in DMN activity and greater synchrony among neural oscillations. These shifts are linked to experiences of ego-dissolution, expanded self-awareness, and transcendental or mystical insights (Brewer et al., 2013; Carhart-Harris et al., 2012).

Theta Waves, Memory, and Intuition

Theta brainwaves are integral to memory encoding and retrieval. They facilitate the consolidation of newly acquired information and are often elevated during learning and problem-solving tasks (Klimesch, 1999). In addition to their cognitive roles, theta waves are associated with intuitive insight and self-awareness. During meditation, heightened theta activity has been linked to improved decision-making and a deeper connection with inner experience (Basar et al., 2001).

Brainwave Changes from Meditation

Meditation is known to increase both alpha (8–12 Hz) and theta (4–8 Hz) wave activity. Alpha waves are characteristic of a calm, alert mental state, while theta waves reflect deeper meditative and intuitive states. The increase in these rhythms contributes to greater emotional balance, relaxation, and attentional control—facilitating a more present-focused awareness (Cahn & Polich, 2006; Tomasino et al., 2012).

Brainwave Frequency

Brain Structure, Intelligence, and Alpha Waves

Cognitive ability is strongly influenced by brain structure. Studies indicate that individuals with greater dendritic density and more complex gray matter arborization tend to perform better on intelligence measures (Genc et al., 2018). Higher levels of alpha wave activity are also associated with mental clarity, working memory performance, and effective problem-solving (Bazanova & Vernon, 2014).

Psychedelics, Entropy, and Mystical States

Psychedelics such as psilocybin and LSD provide insight into how brain activity changes during spiritual or mystical experiences. These compounds reduce activity in the DMN and shift brain oscillations toward more flexible, less predictable patterns—a phenomenon referred to as increased neural entropy (Carhart-Harris et al., 2014). This may explain the heightened imagination, emotional depth, and cognitive flexibility often reported during psychedelic sessions (MacLean et al., 2011; Kometer et al., 2015; Carhart-Harris et al., 2012; Muthukumaraswamy et al., 2013).

Alpha-Theta Neurofeedback and Creativity

Neurofeedback techniques that train individuals to elevate both alpha and theta wave activity have been shown to support creativity and reduce anxiety. By receiving real-time feedback on their brain activity, individuals can learn to shift into more optimal cognitive states that support creative thought and emotional calm (Gruzelier, 2009).

Alpha Waves and Emotional Regulation

Increased alpha brainwave activity has been consistently associated with improved emotional resilience and stability. Individuals with naturally higher alpha activity tend to experience lower levels of stress and enhanced emotional regulation (Davidson, 2004). Mindfulness meditation also plays a key role, helping individuals manage emotional responses and reduce negative affect (Taylor et al., 2011).

Chapter 11: Advanced Theory

A NEW SCIENCE OF AWARENESS

Metacognitive neurophysics describes invariant principles by which awareness interacts with the physical and electromagnetic dynamics of neural systems.

Its axioms and formulas reveal cardinal laws governing the dynamic interplay between consciousness and the material universe.

The first principles, axioms 1-7, frame awareness as a measurable granular phenomenon.

The advanced axioms and corollaries in this chapter codify the mind/matter interface between awareness and neurology.

The resulting theory merges neuroscience and physics into a unified framework for understanding the behavior of awareness and memory through the lens of electromagnetic field dynamics.

The Mind/Matter Interface

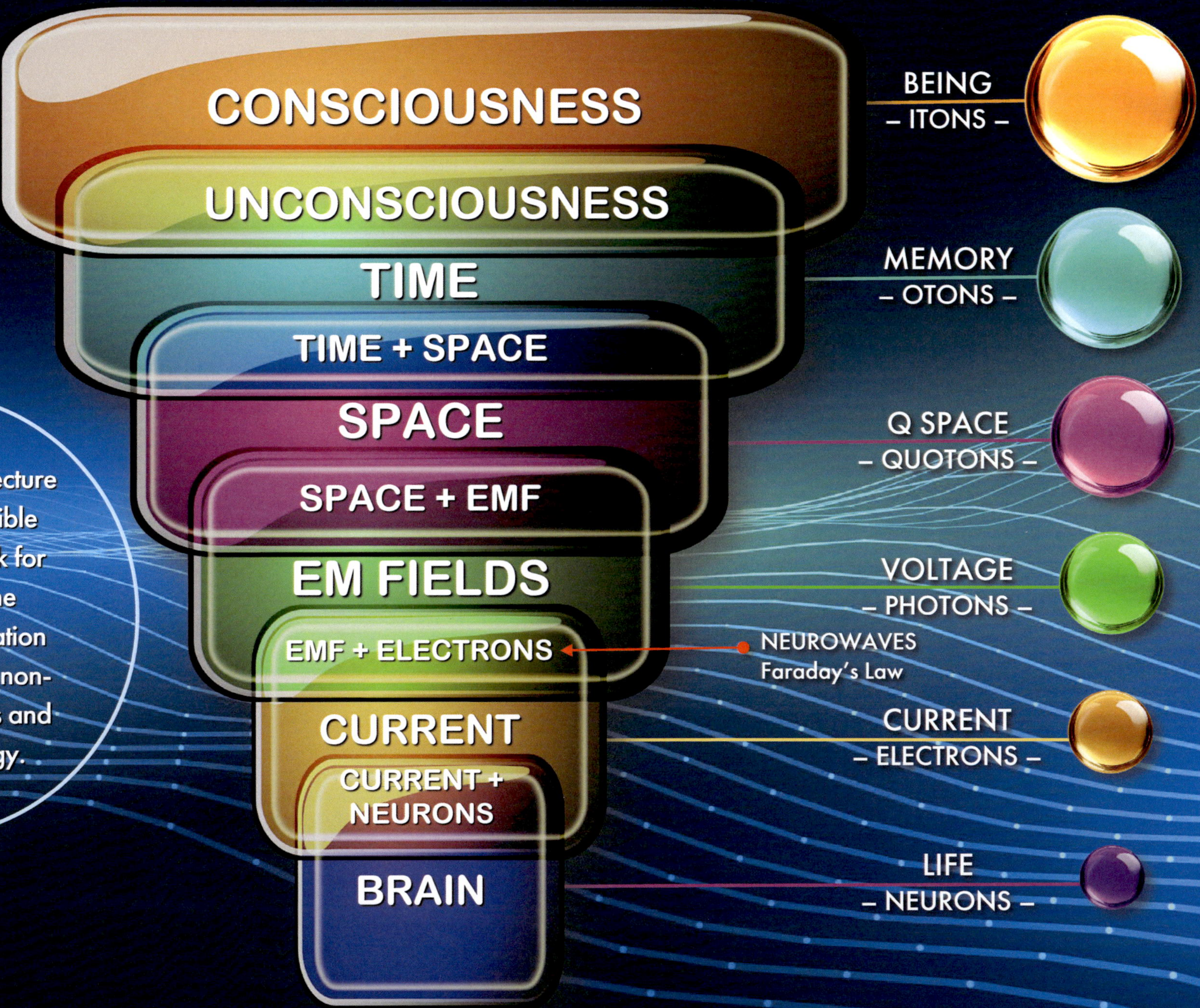

This interface architecture represents a plausible scientific framework for understanding the pathway of information exchange between non-local consciousness and physical neurology.

AXIOM 8:

The Mind/Matter Interface Continuum

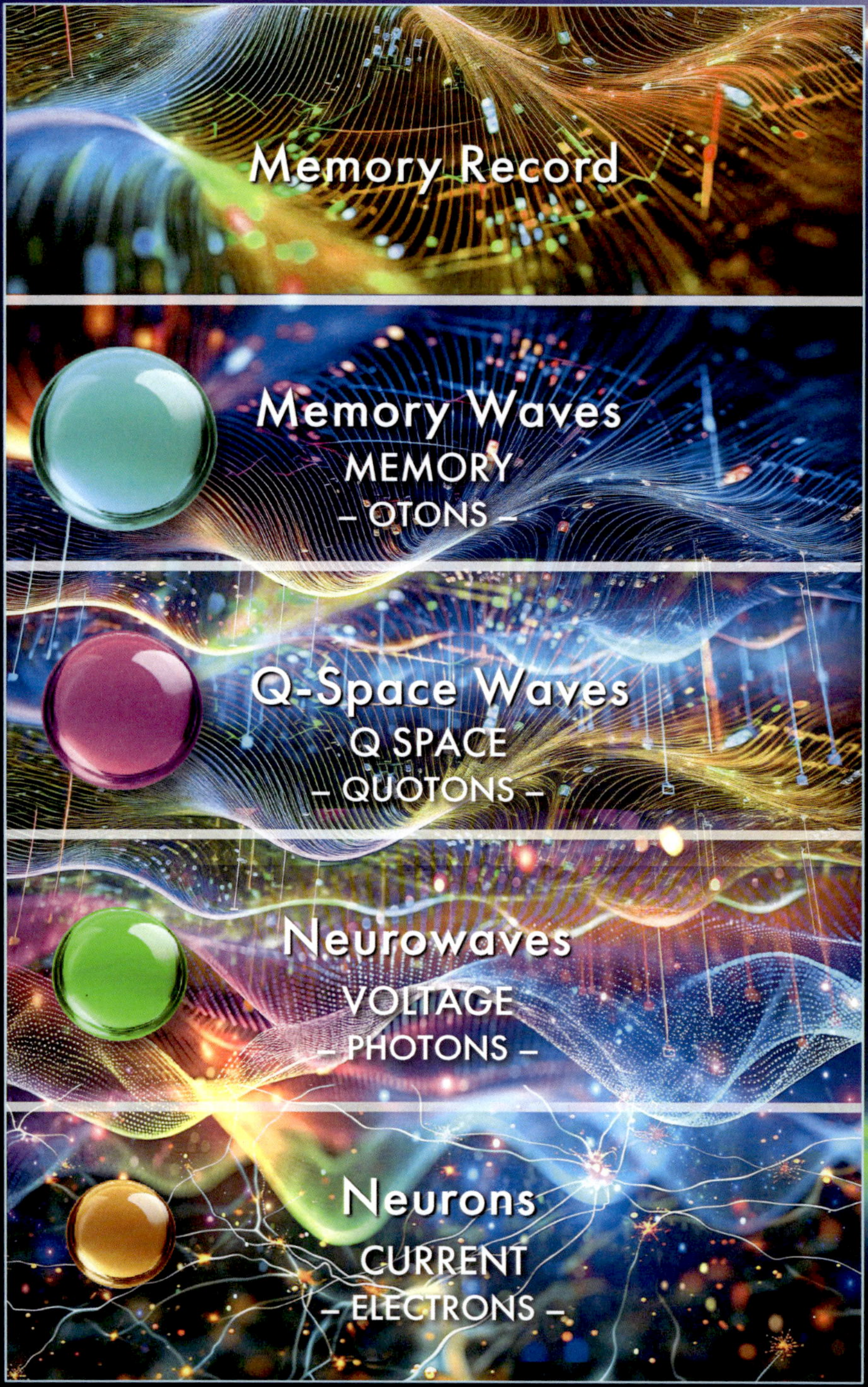

PERCEPTION

5: Memory waves are recorded as new memories. They also activate inert memories that align with the present neural state.

4: Q-Space waves in turn excite memory wave patterns which consciousness can interpret as experience.

3: Neurowaves induce corresponding oscillations in the quantum fabric of space.

2: Neural current generates cellular-level voltage fields that collectively reflect brain states.

1: Incoming perceptions create flowing electric current in neurons.

BEHAVIOR

1: Memory records are activated by intention, thought or perception.

2: Active memory records emit memory waves.

3: Memory waves warp t fabric of quantum space.

4: Q-Space waves in turn generate matching patter neurowave voltage fields.

5: Neurowave voltage propels electrons throug neurons, powering the physical functions of life.

Mind/Matter Interface Elements

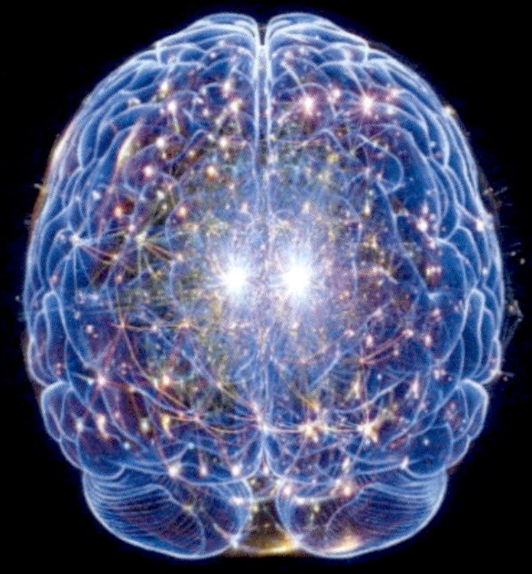

1. MINDPRINTS
Neural activity maps

2. MEMORY RECORDS
Memory pattern storage

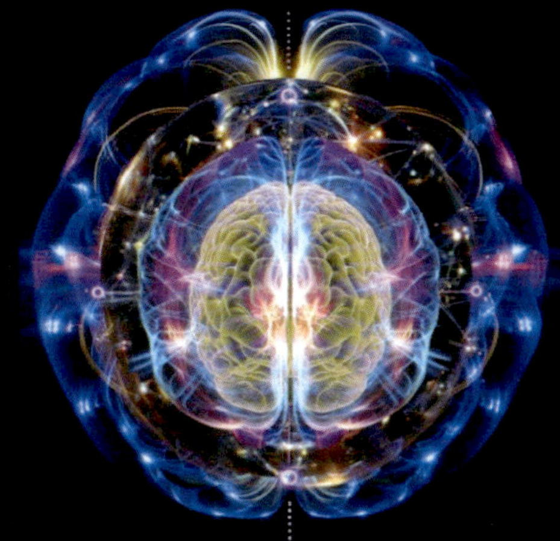

3. MEMORY WAVES
Information-carrying vectors

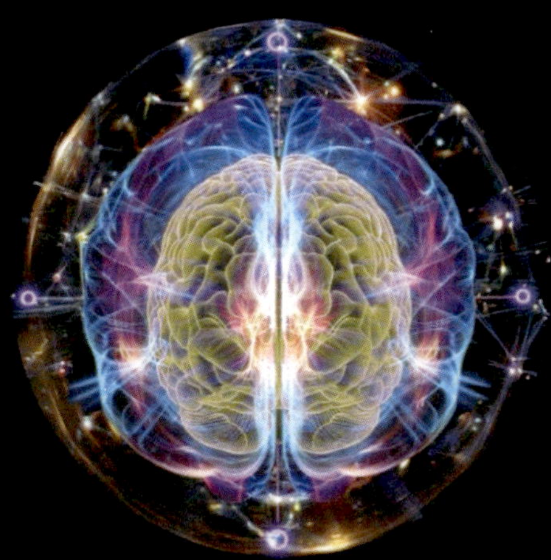

4. Q-SPACE WAVES
Quantum-linking fields

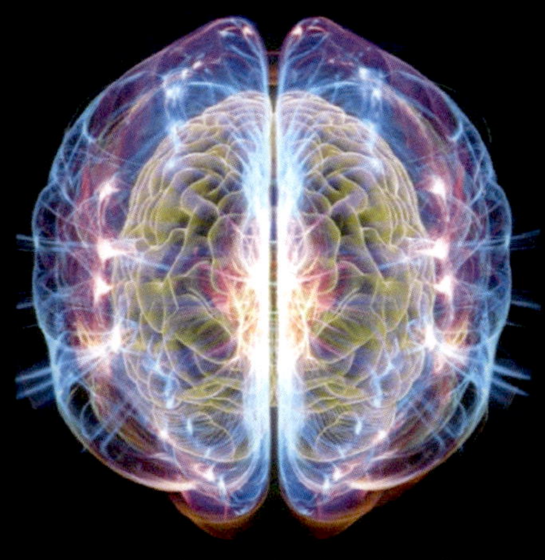

5. NEUROWAVES
Brainwave electromagnetic fields

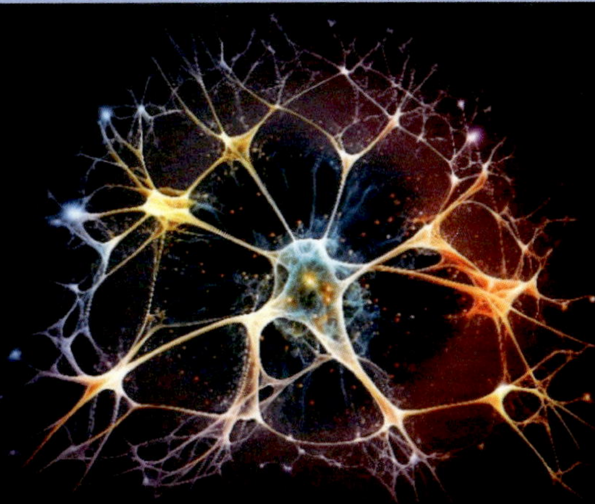

6. NEURONS
Brain's signal processors

1. Mindprint

Mindprint: A multi-dimensional memory trace representing the complete cognitive, emotional, sensory, and spatial imprint of an experience. The mindprint is a snapshot of neural activity patterns, like a brainwave fingerprint.

It is encoded in neural networks and in the geometry of the brain's electromagnetic fields.

A mindprint is reactivatable, context-sensitive, and reflects both the neural substrate and the brainwave field structure. It functions as both an upload and download vehicle—recording experiences and guiding actions.

A memory record contains multiple mindprints covering a specific period of time.

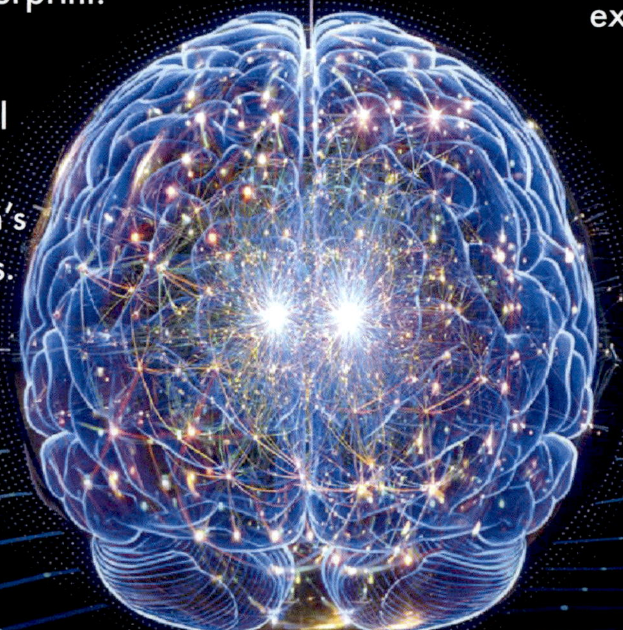

Mindprints propagate through the 4 mind/matter interface layers.

INTERFACE LAYER	MINDPRINT BIT
Memory waves	A peak in a memory wave
Q-Space waves	A peak in a Q-Space wave
Neurowaves	A peak in a neurowave
Neurons	An electrical pulse in a neuron

Mindprint

Behavior: Behavior is based on information retrieved from context-activated memory records; for example, opening a doorknob. The mindprints from these records are downloaded through the mind/matter interface to induce electrodynamic brain activity for generating behavior.

Capacity: A mindprint holds about one gigabyte of neuronal activation pattern data which is streamed at roughly 10 GB per second. (1)

Perception: The mindprint bitstream forms the basis of experience. A human being's innate intelligence translates the bitstream into meaningful perceptual information.

Structure: The mindprint has a three-dimensional holographic structure shaped by the brain's anatomical geometry. Each of its information bits corresponds to a neuron spike somewhere in the brain. Each bit has a unique spatial location in the mindprint based on its originating neuron's position in the brain's 3D geometry.

(1) The brain's 86 billion neurons can be modeled as 86 billion bits of information; roughly 10 gigabytes. About 10% of these circuits are actively signaling at any given time, yielding 1 GB of real-time data and 9 GB of inactivity.

Neurons fire at rates between 4 and 30 Hz depending on brain region and mental state. Using an average rate of 10 Hz, the brain updates its signaling pattern around 10 times per second. This clock speed yields a mindprint real-time data streaming rate of 10 GB per second.

2. Memory Record

A memory record is a non-physical structure that stores *mindprints* – holographic replicas of brainwave configurations from specific moments in time.

Memory records exist outside of time and space in the unconscious awareness field and act as a repository for storing and regenerating experience.

The memory record functions as a multidimensional imprint of neural activity, encoding the frequency, amplitude, and spatial direction of the original brainwave signals.

Memory records are embedded within the mind/matter interface continuum connecting the awareness field to physical neurology.

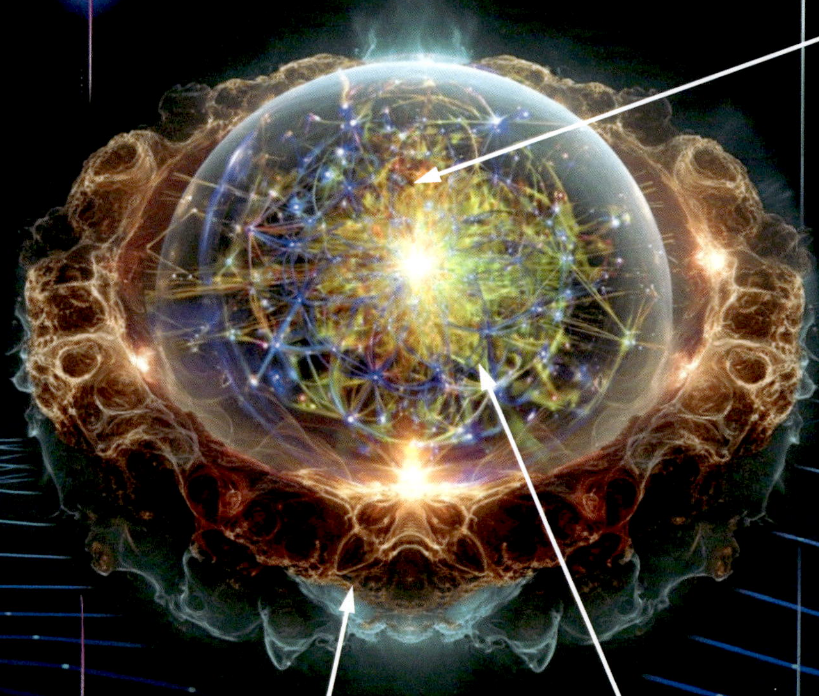

Memory Record

BEHAVIOR

In behavior, memory records initiate action by transmitting stored mindprint patterns back to neurology.

PERCEPTION

In perception, new memory records are continuously formed from the mindprint bitstream conveyed by incoming memory waves.

The number of itons required to activate a memory depends on the amplitude of the memory's recording.

Activating high amplitude memories requires more itons than activating low amplitude memories.

Existing records are also re-excited when resonant frequencies in incoming mindprints match their original encoding.

This reactivates their information content, reanimating its influence on consciousness.

Memory Record Activation

Itons flow through a memory record, acquire its recorded frequencies, and emerge as otons within memory waves.

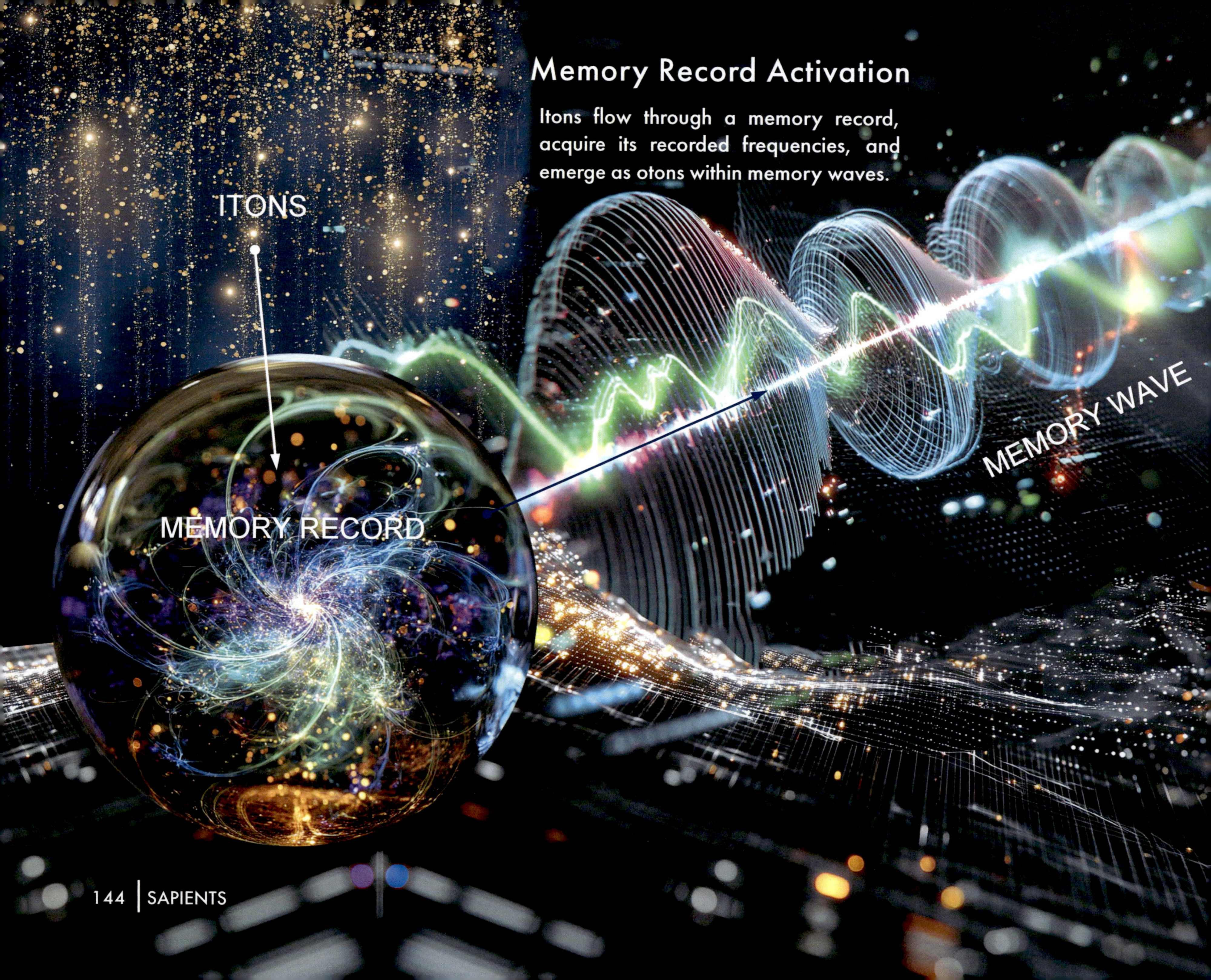

AXIOM 9

A memory has two states: active and inert.

Inert memories store mindprints without drawing awareness. They have no charge; only potential. Almost all memories are inert.

When otons are applied to an inert memory, it becomes active, and reproduces its brainwave recording, which is overlayed on current neurology.

Itons change state to otons when they animate information in a memory. The applied awareness permeates and activates the memory's information, and in this process, becomes unconscious.

When otons are removed from a memory record, the record reverts to its inert state, and the otons return to their original state as itons.

INERT MEMORY

ACTIVE MEMORY

RELATIVE TOTAL VOLUME

Conscious awareness: Marble

Active memory: Tennis ball

Inert memory: Football stadium

3. Memory Waves

Memory waves are non-physical vector fields emitted by memory records in behavior and shaped by Q-Space waves in perception.

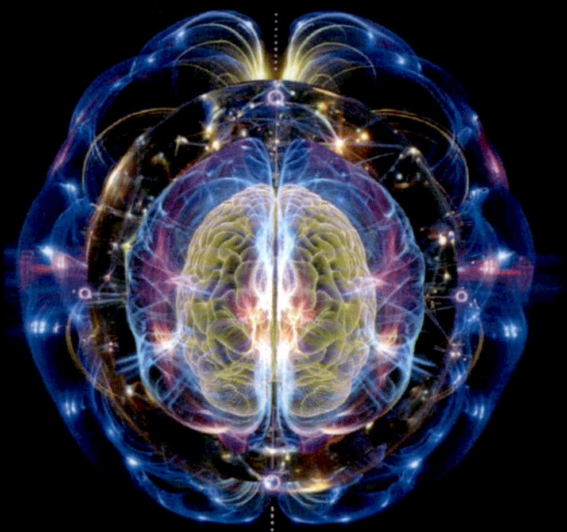

Memory waves are composed of otons.

BEHAVIOR

In behavior, memory waves are the first layer of a cascading wave field stack that connects non-physical awareness to physical brain activity.

They induce oscillations in the fabric of quantum space, producing Q-space waves which, in turn, affect neurowaves in the brain's electromagnetic fields to drive behavior.

PERCEPTION

In perception, memory waves are are the final layer of a perception cascade connecting brain activity to non-physical awareness.

They carry the frequency, amplitude, and directional signature of the originating neurowaves, imprinting this pattern into the awareness field to deliver perception.

4. Q-Space Waves

Quantum-Space waves form a bridge between the non-local memory field and the brain's electromagnetic landscape.

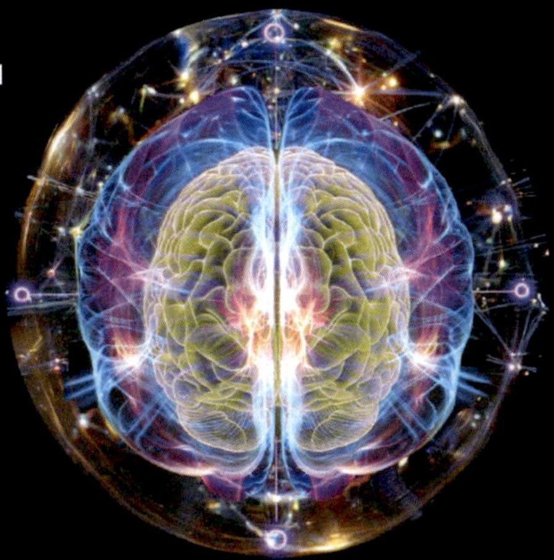

Governed by natural laws, they oscillate and interact dynamically with memory waves above them and neurowaves below them in a closed circuit. They are composed of "quotons."

BEHAVIOR

Q-Space waves are oscillations in the quantum fabric of space generated by memory waves in behavior.

They translate awareness-encoded memory patterns into neurowave fields that can influence the brain's electromagnetic environment.

PERCEPTION

Neurowaves from active brain regions induce oscillations in the fabric of quantum space.

These "Q-Space waves" function as a transducer for uploading neural activity into the domain of awareness, transforming local physical neural activity into non-local quantum information.

5. Neurowaves

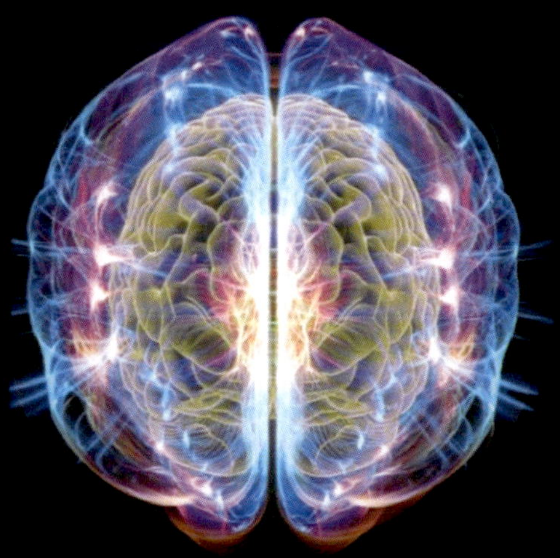

BEHAVIOR

In behavior, electromagnetic neurowave fields in the brain are shaped by incoming Q-Space waves. Neurowaves form the interface layer where Q-Space wave patterns convert into physical neural activity.

Through Faraday's Law, neurowave voltage induces current flow in neurons, driving electrical activity that translates conscious intent or unconscious memory into physiological action.

PERCEPTION

When neurons fire, the current they produce generates electromagnetic waves called neurowaves. (Per Ampere's Law, moving electric currents produce electromagnetic waves.)

These electromagnetic wave patterns carry frequency-encoded information about perceptions and the brain's state.

Neurowaves are composed of photons.

6. Neurons

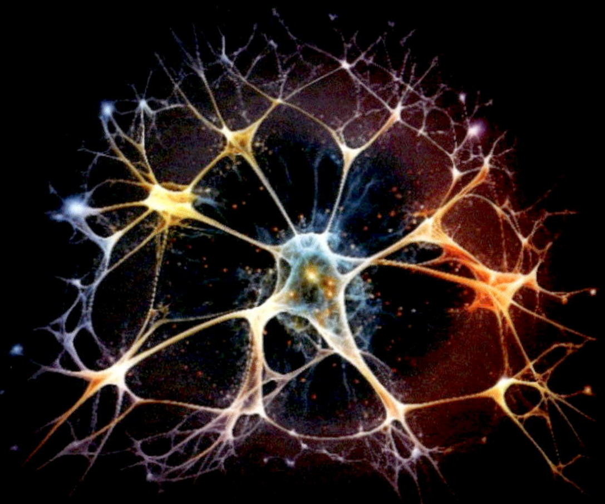

BEHAVIOR

Neurons are the endpoints of the top-down behavioral cascade.

Neurowaves induce electrical currents in neural networks, resulting in neuron firing that produces thought, action, or feeling.

INDUCTION

Faraday's Law

PERCEPTION

Incoming perceptions from the senses drive electric current in the brain.

This electrical activity generates neurowave voltages which initiate the perception chain.

RADIATION

Ampere's Law

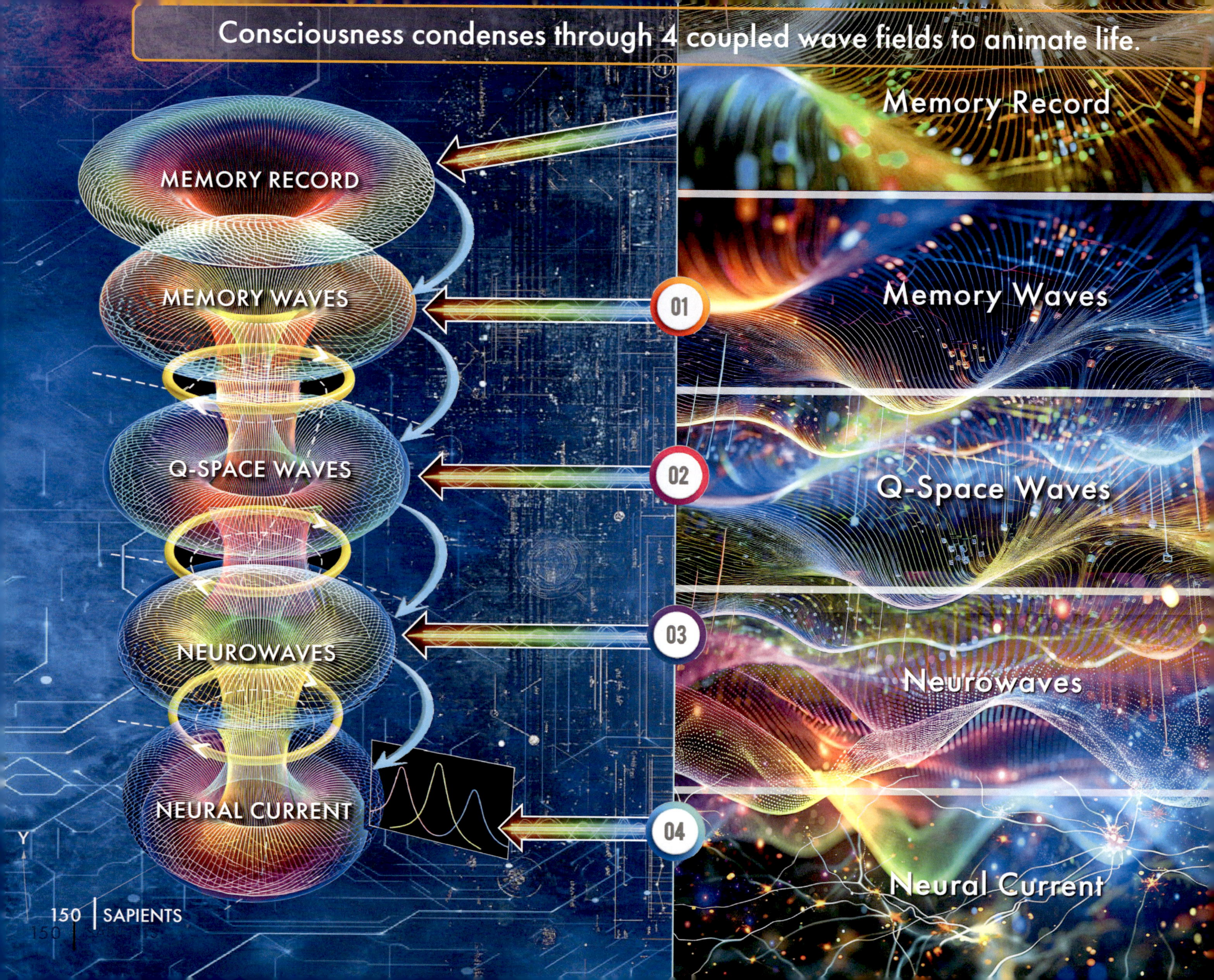
Consciousness condenses through 4 coupled wave fields to animate life.
MEMORY RECORD
MEMORY WAVES
Q-SPACE WAVES
NEUROWAVES
NEURAL CURRENT
01
02
03
04
Memory Record
Memory Waves
Q-Space Waves
Neurowaves
Neural Current

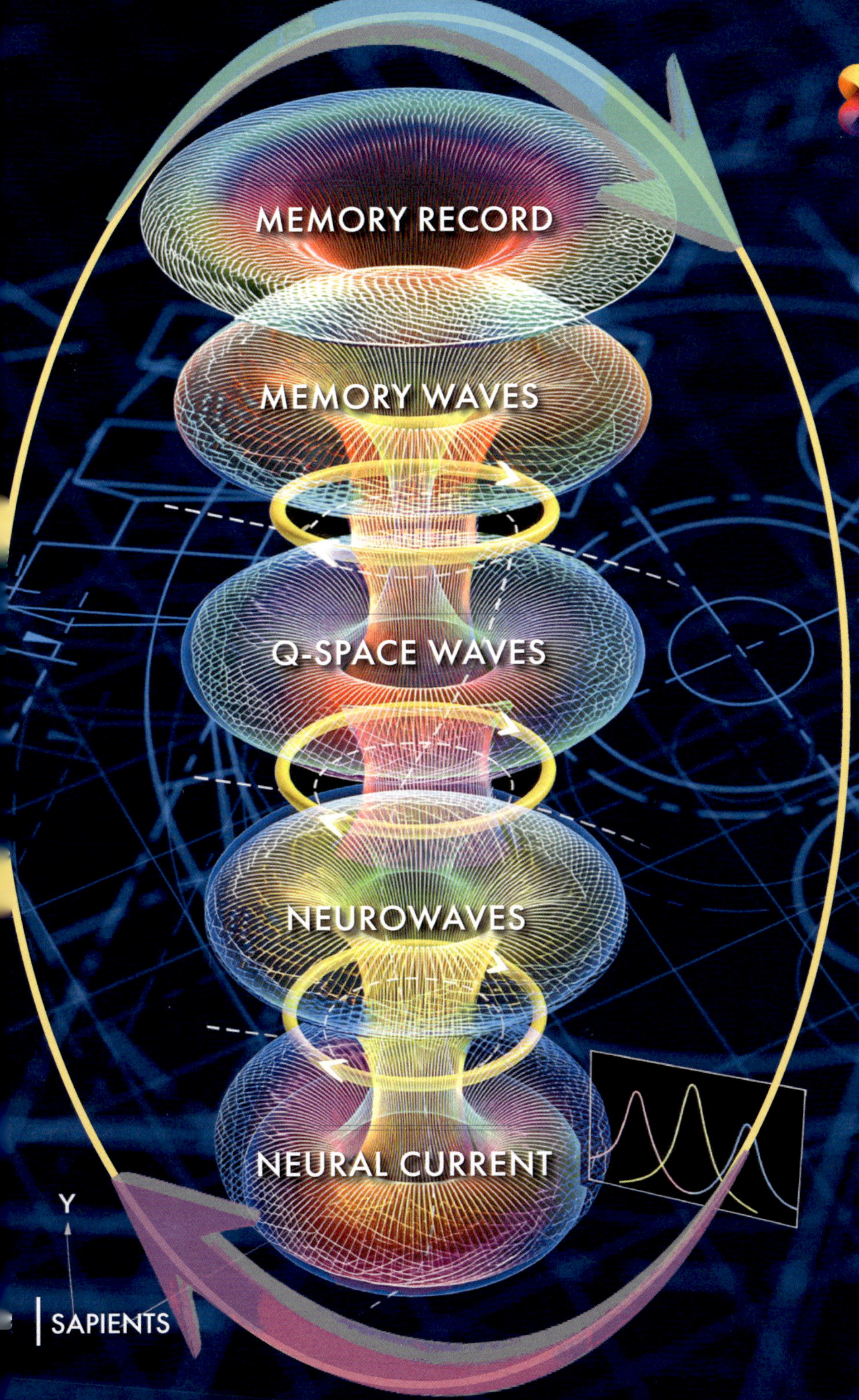

COROLLARY 2
Wave Field Coupled Circuit

A human being's unconscious memory field is joined in a closed "wave field coupled circuit" with their body's neurology.

A wave field coupled circuit is a system where the interaction and transfer of energy and information between circuits is mediated through wave fields.

Wave fields may include magnetic waves along with zero-point energy fields, quantum waves, or other non-local energy fields.

Wave field coupling occurs when overlapping wave fields exchange energy or information.

In humans, this circuit culminates in neurowave fields as the fundamental interface zone between awareness and physical neurology.

Communication across this interface must follow the laws which govern all electromagnetic fields.

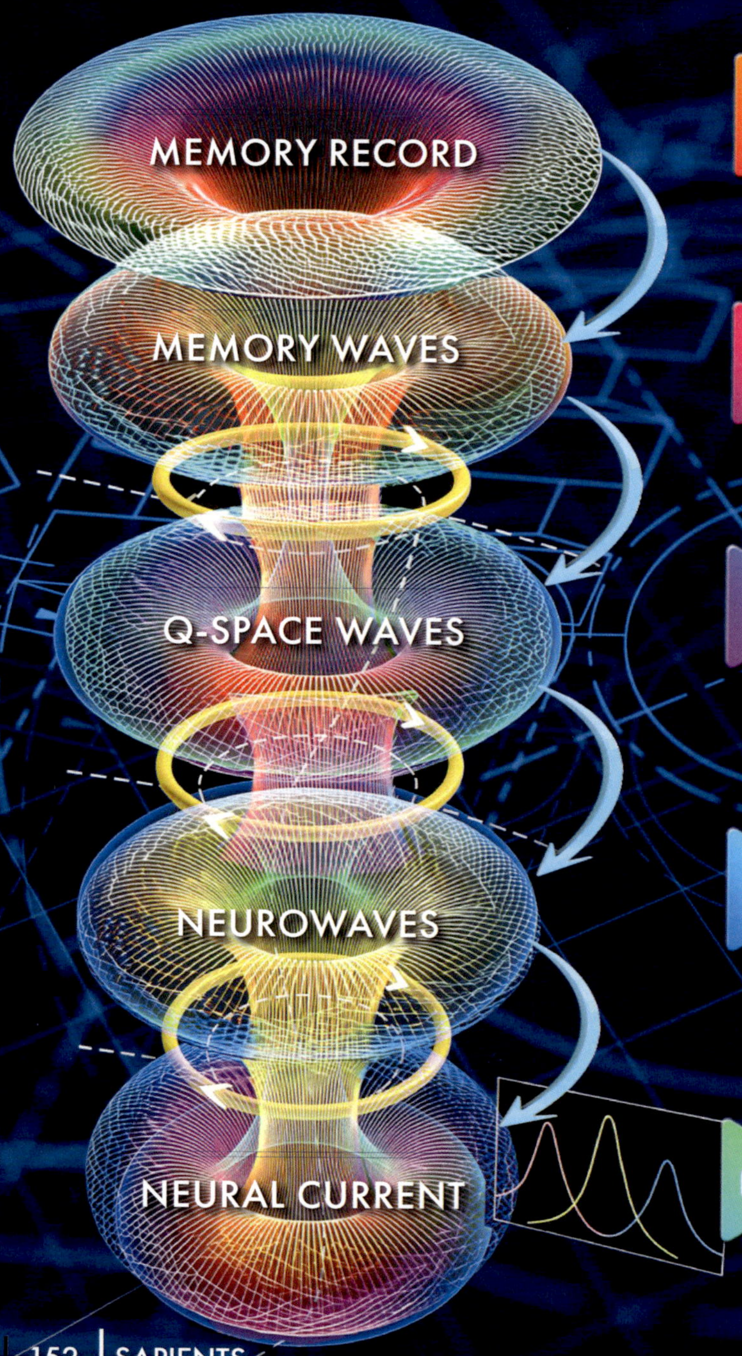

MIND / MATTER INTERFACE – BEHAVIOR

01 **LAYER 1 – Memory Record:** When otons enter an inert memory record, they activate and recharge it. The record's charge is proportional to the number of otons it encloses.

02 **Layer 2 – Memory Waves:** The strength of the memory wave field is proportional to the memory record's charge.

03 **Layer 3 – Q-Space Waves:** The strength of the Q-Space wave vector field is proportional to the enclosing memory wave field charge.

04 **Layer 4 – Neurowaves:** The strength of the neurowave vector field is proportional to the enclosing Q-Space wave field charge.

05 **Layer 5 – Neural Current:** The electrical current passing through a neural network is proportional to the enclosing neurowave field charge.

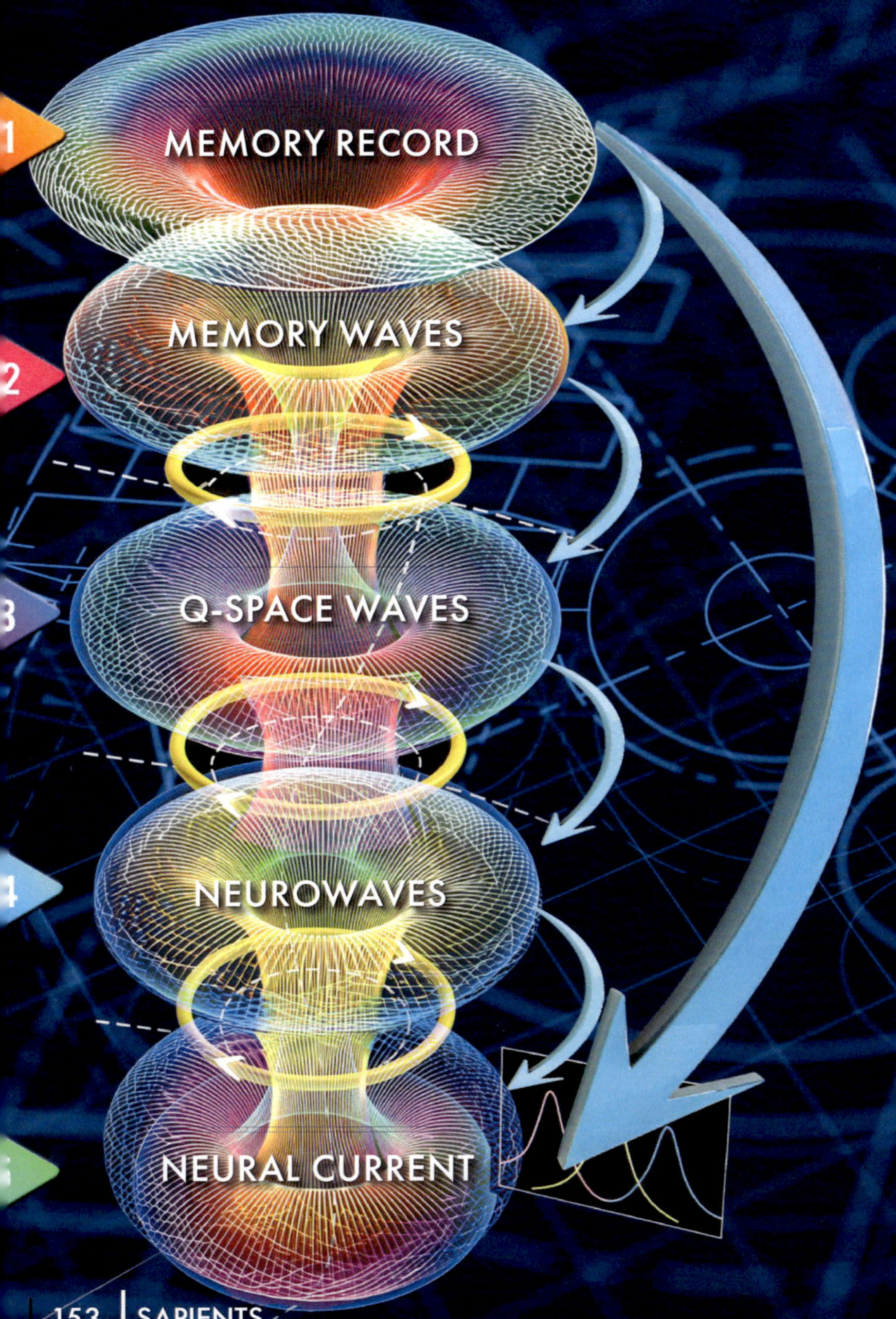

Conflating behavioral layers 1-5:

COROLLARY 3

The strength of current in neural networks associated with memories (Layer 5) is proportional to the number of otons in active memory records (Layer 1) .

COROLLARY 4

Reducing the number of otons in active memory records will cause a corresponding decrease in current flow through the associated neurology.

Increases in free awareness should therefore be accompanied by reductions in brain current.

Empirical evidence for this corollary appears on the following page.

MIND / MATTER INTERFACE – BEHAVIOR

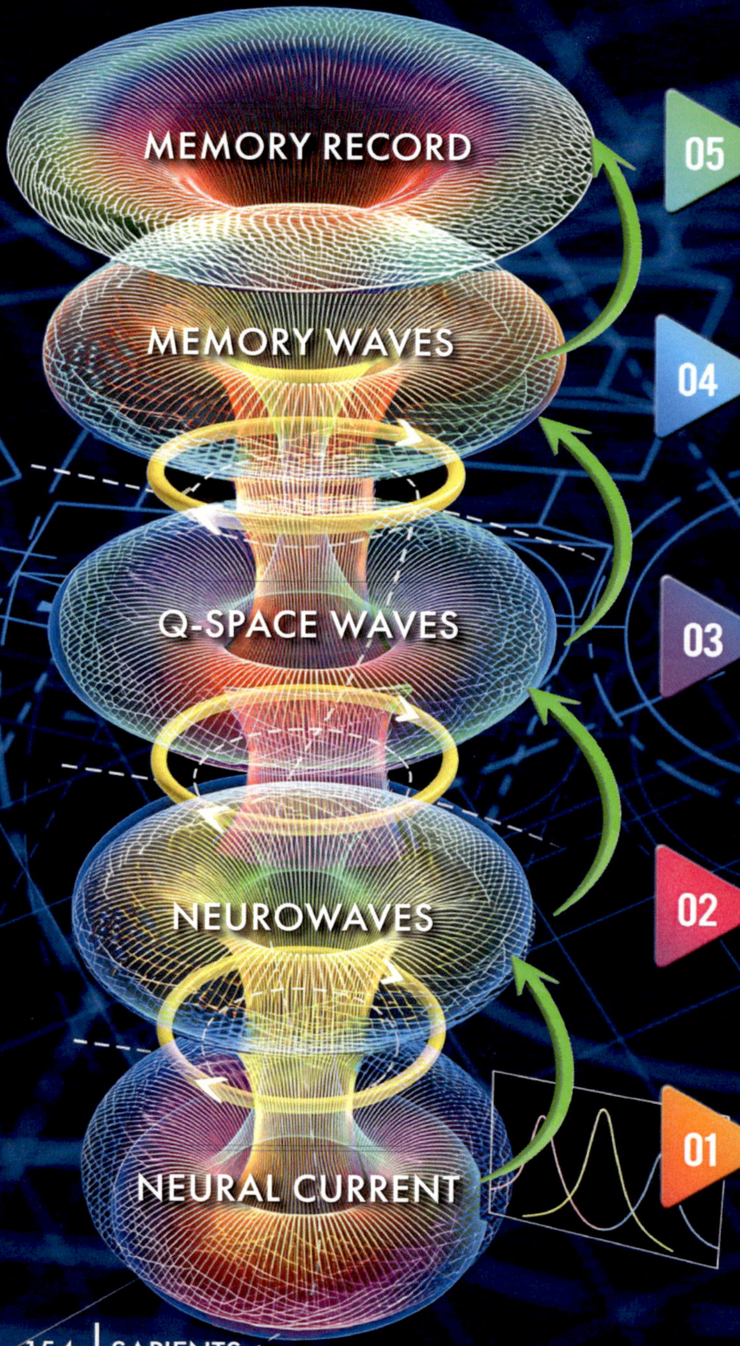

05 LAYER 5 – Memory Record: The memory record's charge is proportional to the strength of the memory waves it registers. The number of otons recruited to animate a memory is proportional to the memory record's charge.

04 Layer 4 – Memory Waves: The strength of the memory wave field is proportional to the enclosing Q-Space wave field charge.

03 Layer 3 – Q-Space Waves: The strength of the Q-Space wave field is proportional to the enclosing neurowave field charge.

02 Layer 2 – Neurowaves: The strength of the neurowave field is proportional to the electrical current passing through a neural network.

01 Layer 1 – Neurons: The electrical current passing through a neural network is proportional to the amplitude of incoming perceptual signals and associated emotional responses.

MIND / MATTER INTERFACE – PERCEPTION

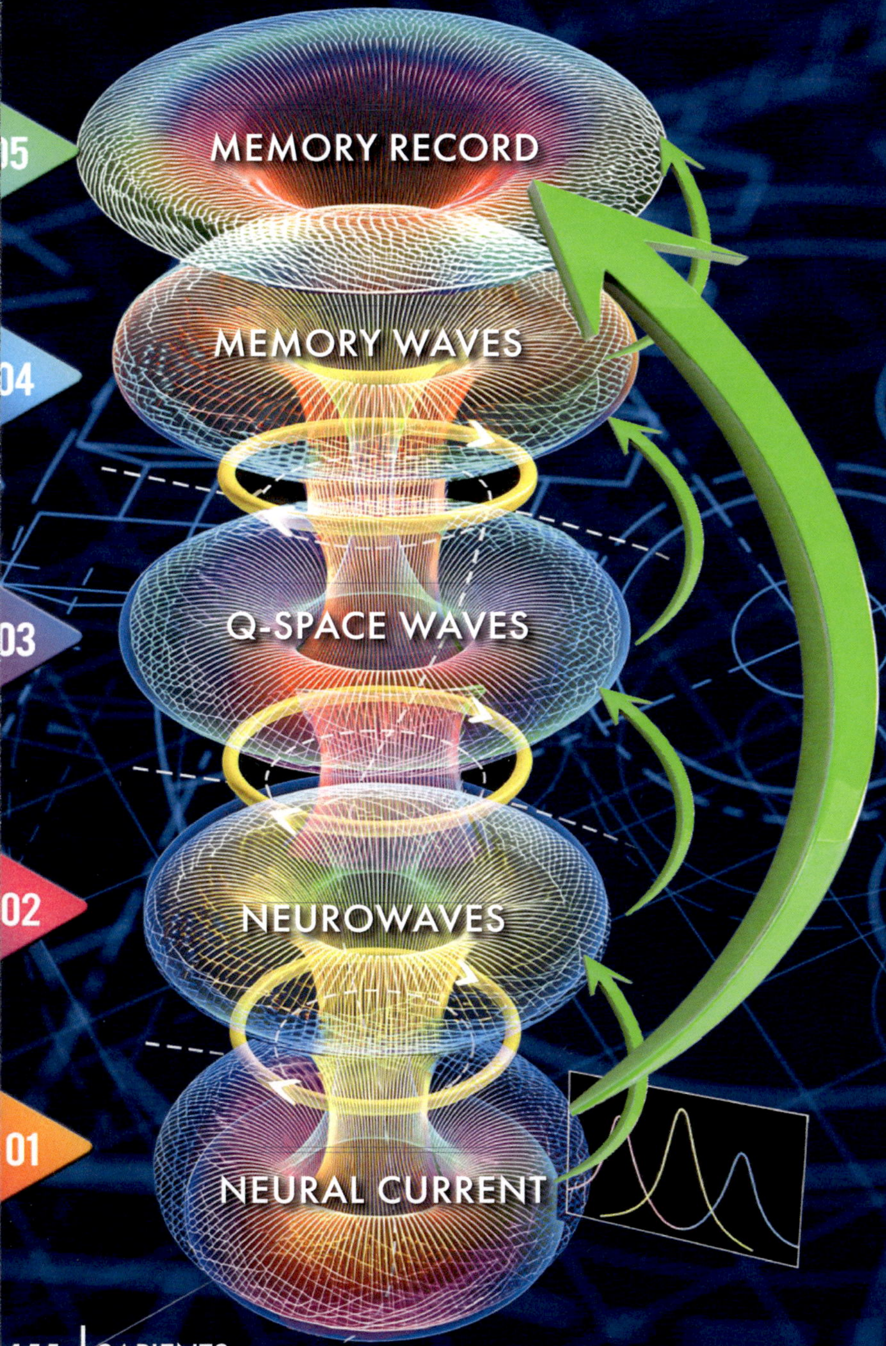

Conflating perceptual layers 1-5:

COROLLARY 5

The number of otons in active memory records (Layer 5) is proportional to the strength of current in the associated neural networks (Layer 1).

COROLLARY 6

Reducing current flow through the neurology associated with a memory will cause a corresponding decrease in the number of otons in the active memory record.

Reductions in brain current should therefore be accompanied by increases in free awareness.

This corollary forms the foundation for the practice of genetic cognitive enhancement.

MIND / MATTER INTERFACE – PERCEPTION

Metacognitive Neurophysics Axioms

1

Awareness is a primary universal element.
Awareness is fundamental and irreducible. It is a foundational principle.

2

Awareness is composed of discrete packets.
In the same way that light is made of photons, awareness consists of elemental units.

3

Awareness packets have two states.
itons: conscious awareness packets
otons: unconscious awareness packets

4

Awareness can change state.
Awareness packets can change state between conscious and unconscious.

5

Free awareness is conscious and formless.
Conscious awareness packets have no amplitude or wavelength.

6

Applied awareness takes form and becomes unconscious.
Unconscious awareness takes the form of information.

Metacognitive Neurophysics Axioms

7
Conscious awareness is reciprocal with brainwave amplitude and neural activity.

8.1 PERCEPTION
Mind/Matter Interface Continuum
Neural activity rises through five layers to reach conscious awareness.

8.2 BEHAVIOR
Mind/Matter Interface Continuum
Consciousness condenses through five layers to form the living universe.

9
A memory has two states: active and inert.
Inert memories store information without drawing awareness.

*****
When itons are applied to an inert memory, they change state to otons, energizing the memory's recorded information.

*****
When otons exit an active memory, it returns to its inert state, and the otons revert to itons.

Metacognitive Neurophysics Corollaries

1 Conscious awareness packets have no amplitude or wavelength.

2 A human being's unconscious mem field is joined in a closed "wave fiel coupled circuit" with their body's neurology.

3 The strength of current in neural networks associated with memories is proportional to the number of otons in active memory records.

BEHAVIOR

4 The number of otons in active memory records is proportional to the strength of current in the associated neural networks.

PERCEPTION

PERCEPTION

BEHAVIOR

5 Reducing the number of otons in active memory records will cause a corresponding decrease in current flow through the associated neurology.

BEHAVIOR

6 Reducing current flow through the neurology associated with a memory will cause a corresponding decrease i the number of otons in the active memory record.

PERCEPTION

Historical Precedents

AXIOM 1

Awareness is a primary universal element. Awareness is fundamental and irreducible.

1803: John Dalton proposed that matter is made of atoms. He had no microscope or direct evidence. His atomic theory was proven in 1908 and forms the foundation of chemistry.

AXIOM 2

Awareness is composed of discrete packets. In the same way that light is made of photons, awareness consists of elemental units.

1897: J.J. Thomson identified the electron — not by seeing it, but by observing cathode rays. He inferred what could not yet be seen. Inference first, evidence later.

AXIOM 4

Awareness can change state. Awareness packets can alternate between conscious and unconscious form.

Isaac Newton described the laws of gravity in 1687. Centuries later, Einstein used calculus to prove gravity, but Newton discovered gravity, and invented calculus.

AXIOM 8

Mind/Matter Interface Continuum Consciousness condenses through five layers to form the living universe.

Science has always invited us to believe in what cannot yet be seen — atoms, electrons, gravity.

Metacognitive neurophysics is simply a new iteration in this evolutionary process.

Its axioms may seem speculative to some today, but like Dalton's atoms, they provide a framework for research.

As instruments catch up and evidence grows, the axioms may someday be widely regarded as the cornerstone of a new science.

Chapter 12: Formulas

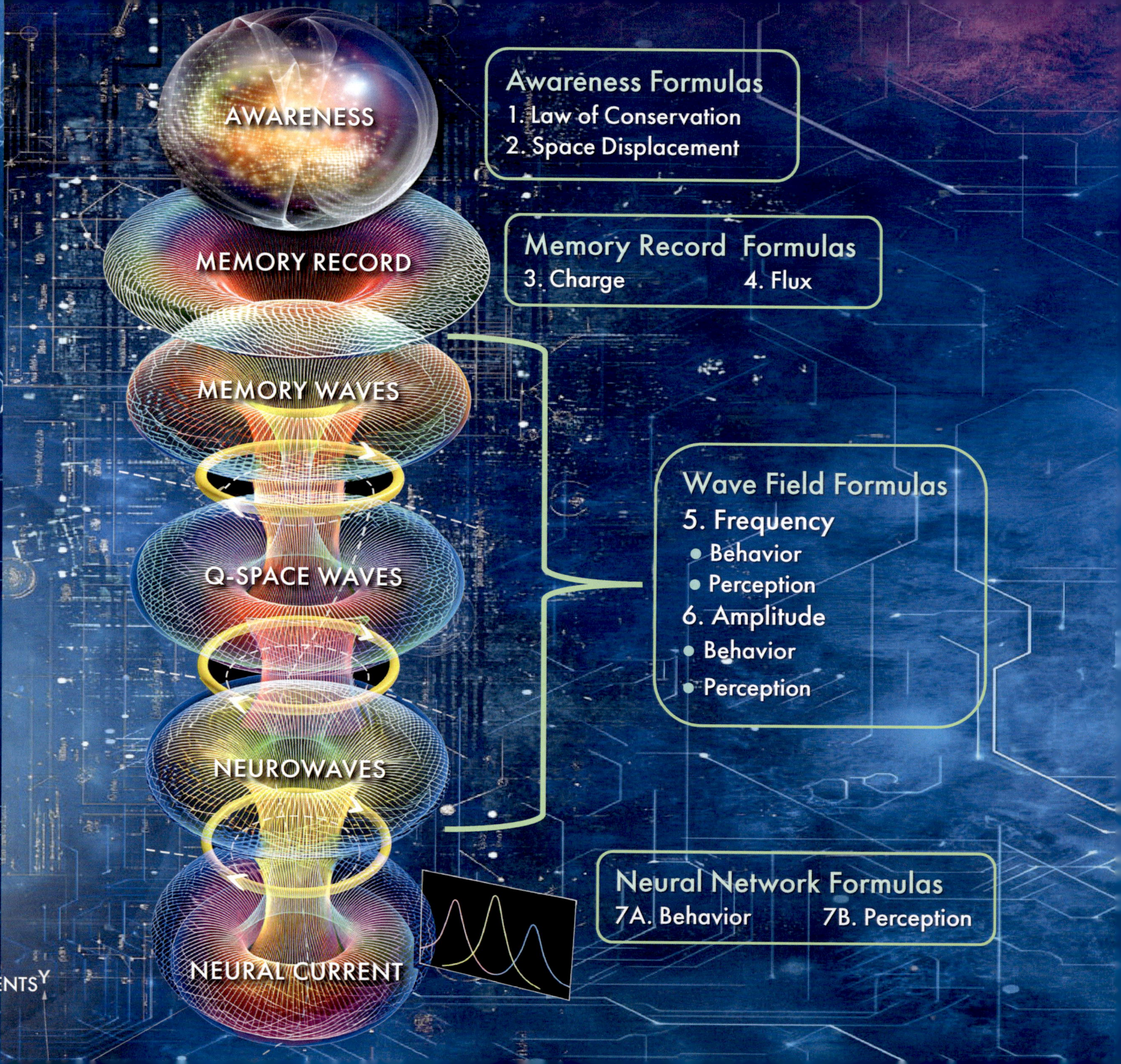

The Law of Conservation of Awareness

Awareness cannot be created or destroyed;
only changed from one form to another.

1

$$I = (n - O)$$

where:
I = conscious awareness packets (itons)
O = unconscious awareness packets (otons)
n = total number of awareness packets per individual
and where:
n = I + O

The measurements of consciousness and unconsciousness are reciprocals of one another; as one increases, the other decreases. However, an individual's total amount of awareness remains constant, reflecting a law of conservation.

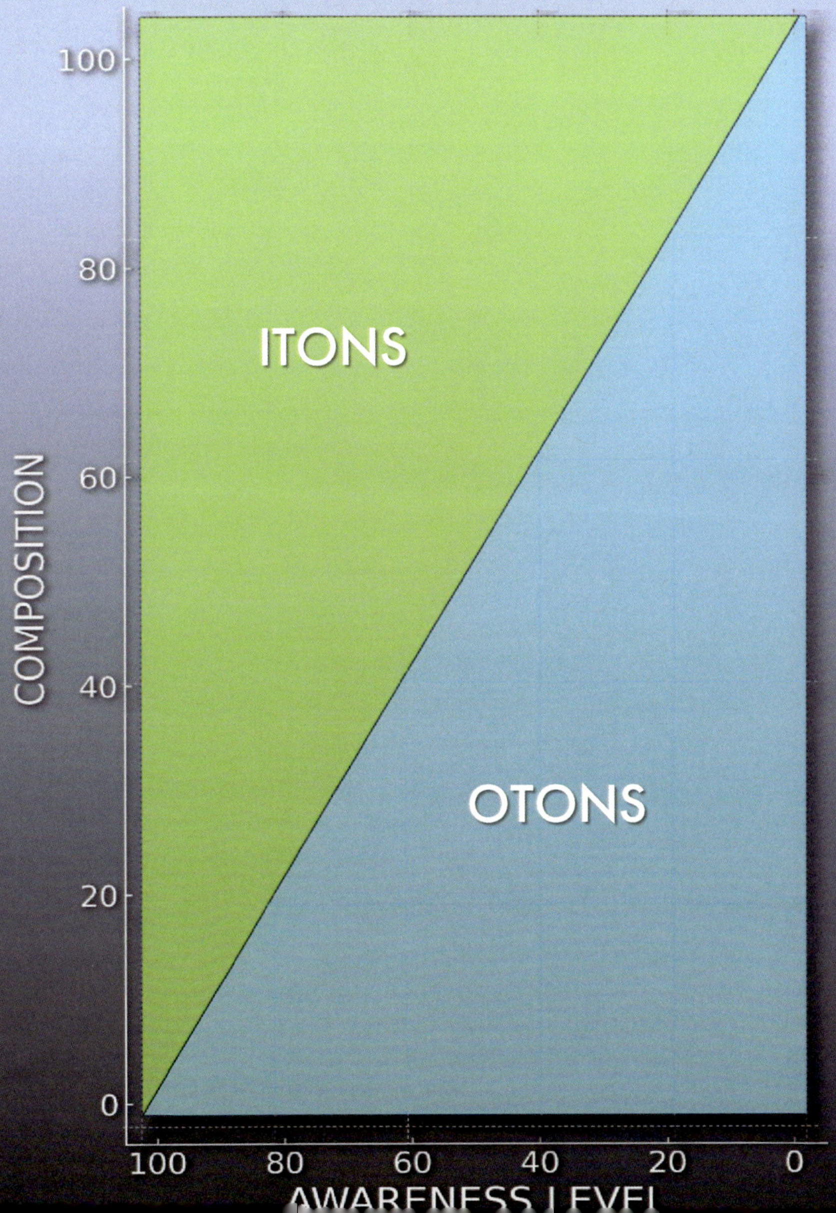

The Space Displacement Formula

The volume of space displaced by an awareness field is proportional to the number of itons in the field.

2

$$V = \frac{4}{3}\pi \, \mathrm{i}^{2}$$

where:

V = volume of space displaced
i = number of itons in an awareness field

Itons displace space like air bubbles in water.

Otons create the living universe by occupying memory, space, energy and matter.

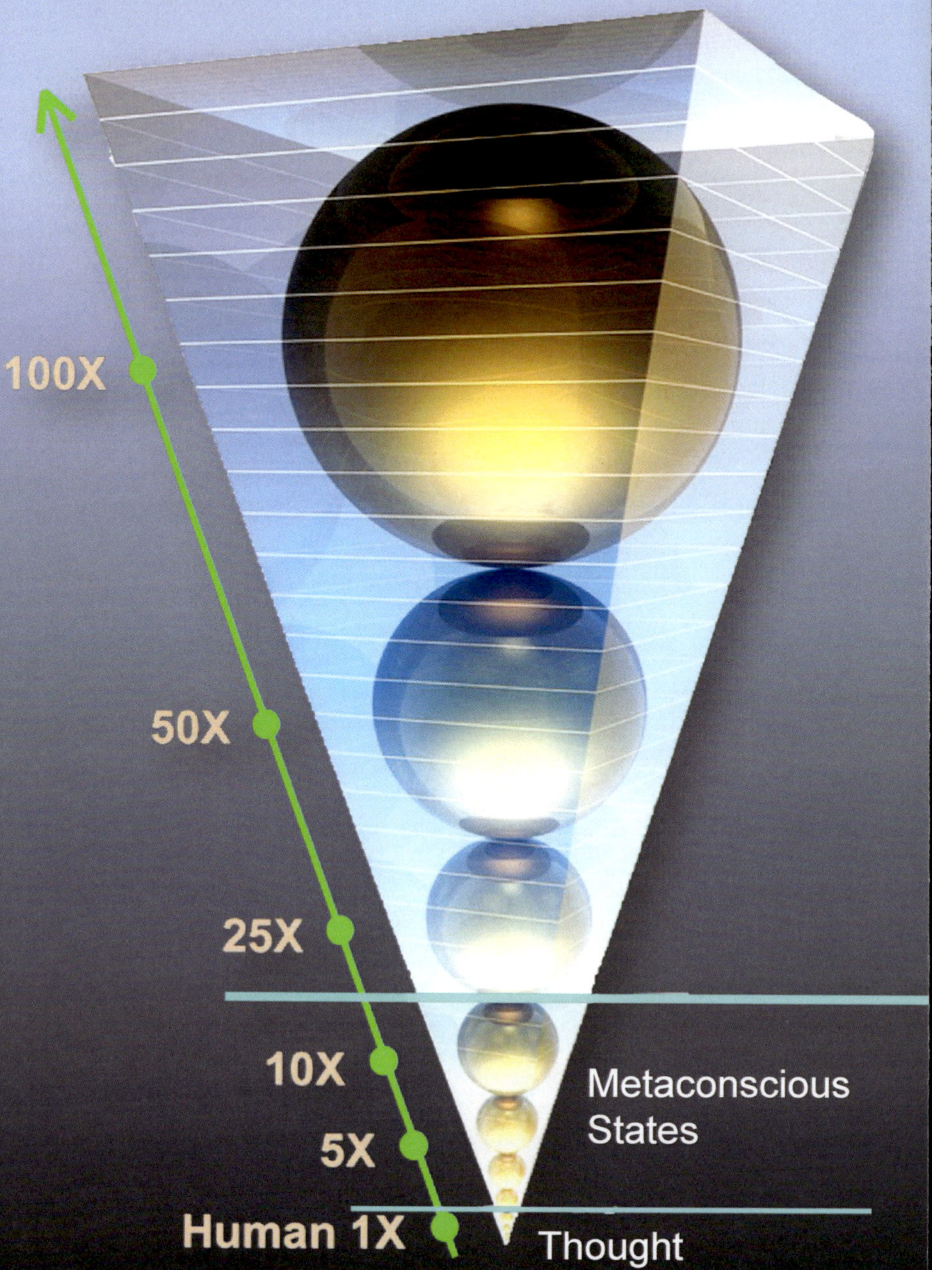

Memory Record Charge

3

The amplitude of a memory's recording depends on the strength of its originating electrical signals in neurology.

Higher amplitude memories have more charge. Charge is the differential between a memory record's plus and minus end points (source and sink).

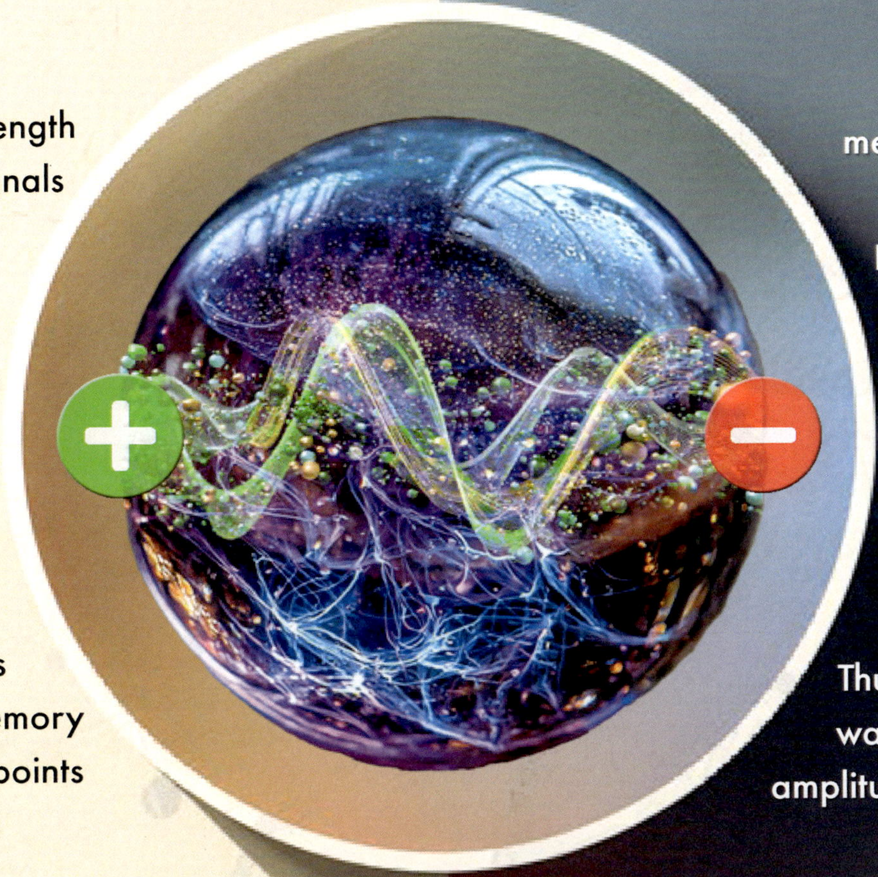

The strength of an activated memory's polarity (plus/minus difference) is the force of its projection through the closed wave field coupled circuit.

Thus, the force or voltage of the wave field circuit varies with the amplitude of an activated memory's recording.

Memory Record Polarity

A memory must have plus and minus end points (source and sink) in order to oscillate.

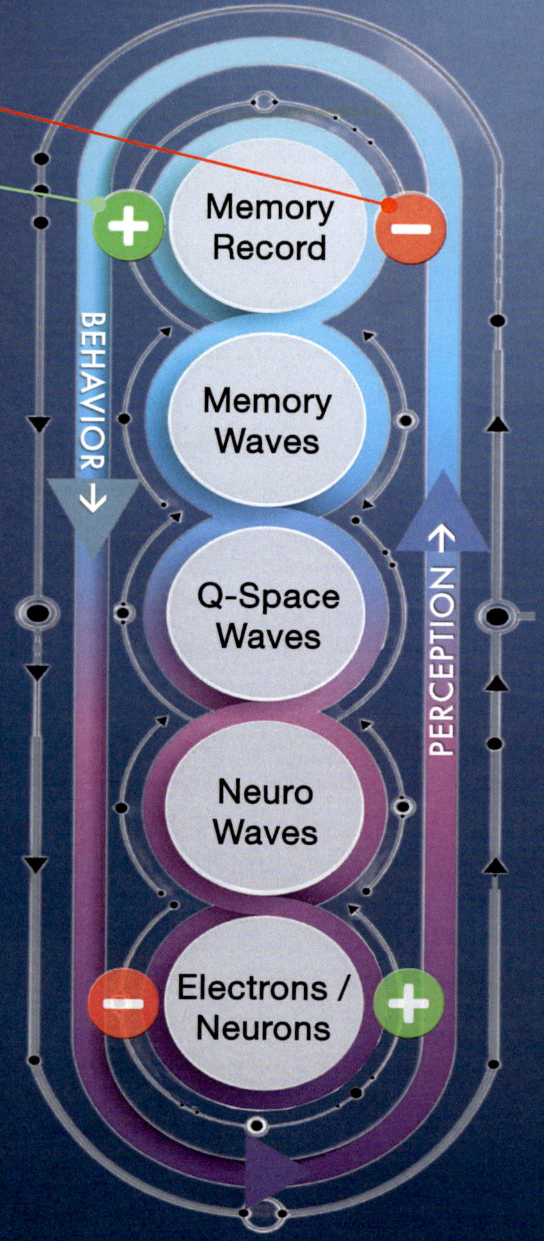

Memories are recordings of oscillating neurowave signatures in neurology. Since these waves have polarity, their replicas also have polarity.

The memory's polarity drives energy through a closed wave field coupled circuit from the plus to the minus terminal of the memory.

The Memory Record Charge Formula

A memory record's charge is proportional to the total number of otons enclosed by the record.

$\oint$: denotes the closed surface integral (1)

$d\mathbf{a}$: the infinitesimal area vector on the closed surface, pointing outward (2)

$\epsilon 0$: a permittivity constant defining how otons enter a memory record

$$\oint \mathbf{MR} \cdot\cdot\text{-}\, d\mathbf{a} = Q\text{enc} \,/\, \epsilon 0$$

MR: the memory record charge

Qenc: the number of otons enclosed by the memory record

(1) A closed surface integral ($\oint$) is a calculus tool used to sum up a quantity over a closed surface.

(2) A differential area element (da) is a calculus tool used to break down complex surfaces into manageable, infinitesimally small pieces for integration purposes.

4

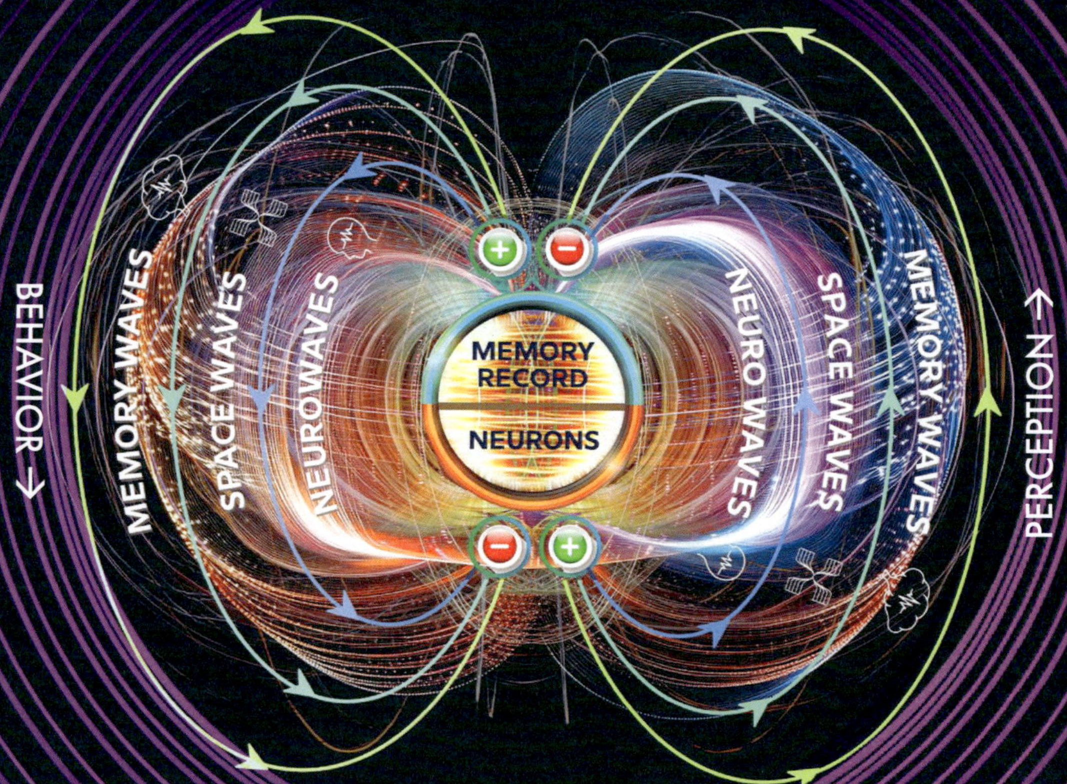

Wave Field Circuit Flux

The flux through any memory wave field coupled neural circuit is zero.

$$\nabla \cdot \mathbf{M} = 0$$

Where:

∇ is the divergence operator

M is the memory field

No waves diverge from or flow out of the closed memory wave field circuit. It is self-contained.

The wave field coupled circuit is technically a dipole, meaning it has two poles; plus and minus. There are no memory monopoles.

166 SAPIENTS

5

Laws of Behavior and Perception

Wave Field Formulas

Frequency

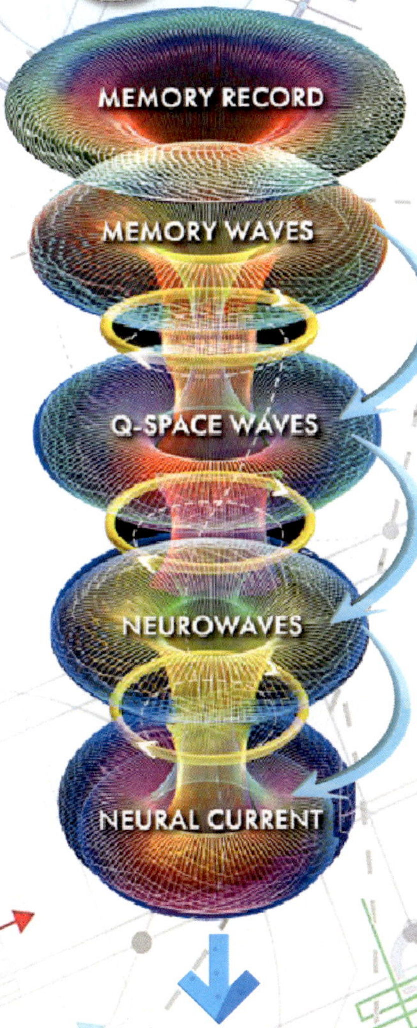

Behavior

5A. A time-varying memory field induces a time-varying space field.

5B A time-varying space field induces a time-varying neurowave field.

5C. A time-varying neurowave field induces a time-varying electrical field.

5F. A time-varying space field induces a time-varying memory field.

5E. A time-varying neurowave field induces a time-varying space field.

5D A time-varying electrical field induces a time-varying neurowave field.

Perception

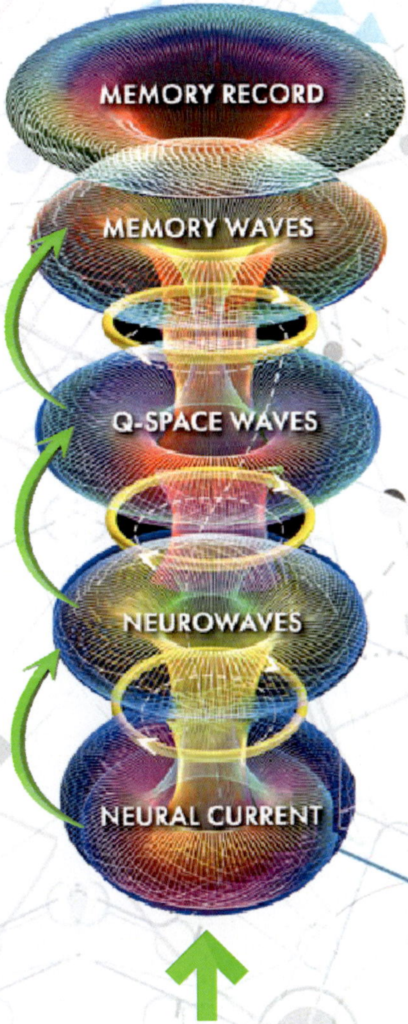

Wave Field Frequency – Behavior

5A. Memory/Space Wave Field Frequency

A time-varying memory field induces a time-varying space field.

$$\frac{\partial M}{\partial t} k = \frac{\partial S}{\partial t}$$

M is the memory wave field
S is the space wave field
t is time

∂M/∂t is the time derivative of field M, representing how M changes over time.

∂S/∂t is the time derivative of field S, representing how S changes over time.

5B. Space/Neurowave Field Frequency

A time-varying space field induces a time-varying neurowave field.

$$\frac{\partial S}{\partial t} k = \frac{\partial N}{\partial t}$$

S is the space wave field
N is the neurowave field
t is time

∂S/∂t is the time derivative of field S, representing how S changes over time.

∂N/∂t is the time derivative of field N, representing how N changes over time.

5C. Neurowave/Current Field Frequency

A time-varying neurowave field induces a time-varying electrical field.

$$\frac{\partial N}{\partial t} k = \frac{\partial E}{\partial t}$$

N is the neurowave field
E is the electrical field
t is time

∂N/∂t is the time derivative of field N, representing how N changes over time.

∂E/∂t is the time derivative of field E, representing how E changes over time.

∂ — ∂ is a partial derivative taken for one variable (time *t*) while holding other variables constant.

k — k is a proportionality constant relating the change rate of the first field to the change rate of the second field.

Wave Field Frequency – Perception

5D. Current/Neurowave Field Frequency

A time-varying electrical field induces a time-varying neurowave field.

$$\frac{\partial E}{\partial t} k = \frac{\partial N}{\partial t}$$

E is the electrical field
N is the neurowave field
t is time

∂E/∂t is the time derivative of field E, representing how E changes over time.

∂N/∂t is the time derivative of field N, representing how N changes over time.

5E. Neurowave/Space Field Frequency

A time-varying neurowave field induces a time-varying space field.

$$\frac{\partial N}{\partial t} k = \frac{\partial S}{\partial t}$$

N is the neurowave field
S is the space wave field
t is time

∂N/∂t is the time derivative of field N, representing how N changes over time.

∂S/∂t is the time derivative of field S, representing how S changes over time.

5F. Space/Memory Wave Field Frequency

A time-varying space field induces a time-varying memory field.

$$\frac{\partial S}{\partial t} k = \frac{\partial M}{\partial t}$$

S is the space wave field
M is the memory field
t is time

∂S/∂t is the time derivative of field S, representing how S changes over time.

∂M/∂t is the time derivative of field M, representing how M changes over time.

∂ — ∂ is a partial derivative taken for one variable (time t) while holding other variables constant.

k — k is a proportionality constant relating the change rate of the first field to the change rate of the second field.

Laws of Behavior and Perception

Wave Field Formulas

Amplitude

6

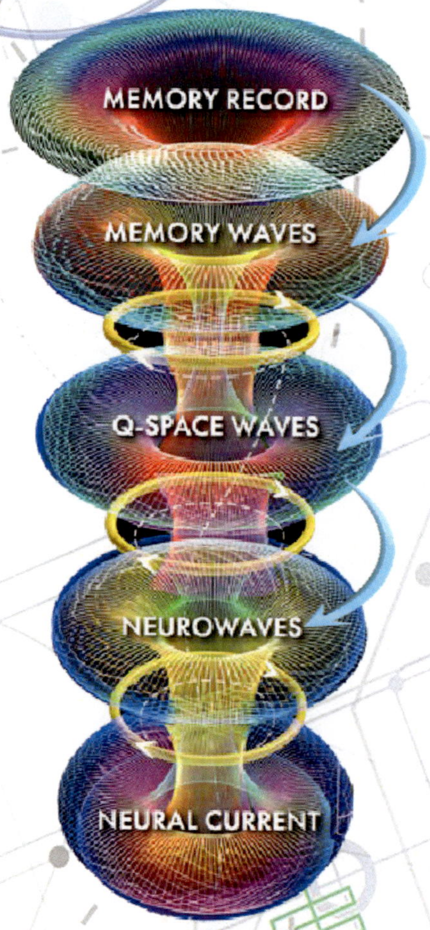

Behavior

6A. The strength of the memory wave field is proportional to the enclosing memory record charge.

6B. The strength of the zero point space field is proportional to the enclosing memory wave field charge.

6C. The strength of the neurowave field is proportional to the enclosing space wave field charge.

Perception

6D. The strength of the space wave field is proportional to the enclosing neurowave field charge.

6E. The strength of the memory wave field is proportional to the enclosing space wave field charge.

6F. A memory record's charge is proportional to the enclosing memory wave field charge.

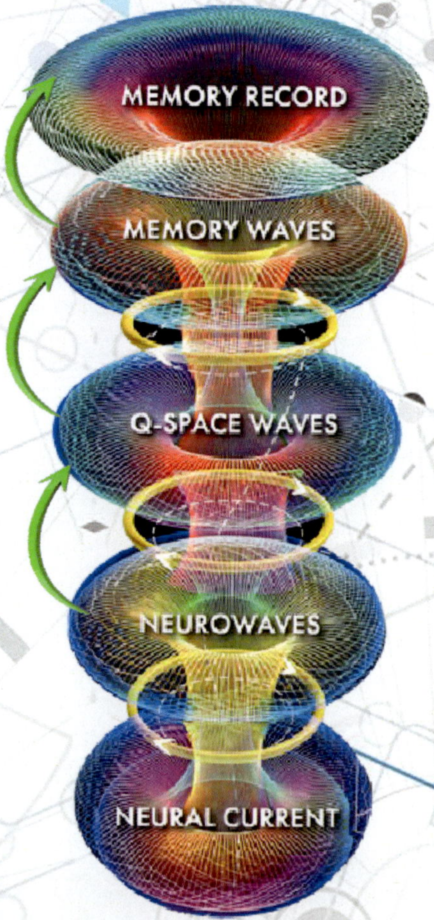

Wave Field Amplitude – Behavior

6A. Memory Wave Field Amplitude

memory wave field's strength is roportional to the enclosing memory record charge.

$$\oint M \cdot \text{-} da = Qenc_{MR} / \epsilon 0_M$$

$\oint M$ is the closed surface integral of the memory wave vector field

da is the infinitesimal area vector on the closed surface, pointing outward

$Qenc_{MR}$ is the total charge enclosed by the memory record's surface MR

$\epsilon 0_M$ defines how much memory wave flux can pass through a memory record

6B. Space Wave Field Amplitude

The strength of the zero point space field is proportional to the enclosing memory wave field charge.

$$\oint S \cdot \text{-} da = Qenc_M / \epsilon 0_S$$

$\oint S$ is the closed surface integral of the space wave vector field

da is the infinitesimal area vector on the closed surface, pointing outward

$Qenc_M$ is the total charge enclosed by the memory wave field's surface M

$\epsilon 0_S$ defines how much memory wave flux can pass through a space wave field

6C. Neurowave Field Amplitude

The strength of the neurowave field is proportional to the enclosing space wave field charge.

$$\oint N \cdot \text{-} da = Qenc_S / \epsilon 0_N$$

$\oint N$ is the closed surface integral of the neurowave vector field

da is the infinitesimal area vector on the closed surface, pointing outward

$Qenc_S$ is the total charge enclosed by the space wave field's surface S

$\epsilon 0_N$ defines how much space wave flux can pass through a neurowave field

Wave Field Amplitude – Perception

6D. Space Wave Field Amplitude

The strength of the space wave field is proportional to the enclosing neurowave field charge.

$$\oint S \cdot -da = Qenc_N / \epsilon 0_S$$

$\oint S$ is the closed surface integral of the space wave vector field

da is the infinitesimal area vector on the closed surface, pointing outward

$Qenc_N$ is the total charge enclosed by the neurowave field's surface N

$\epsilon 0_S$ defines how much neurowave flux can pass through a space wave field

6E. Memory Wave Field Amplitude

The strength of the memory wave field is proportional to the enclosing space wave field charge.

$$\oint M \cdot -da = Qenc_S / \epsilon 0_M$$

$\oint M$ is the closed surface integral of the memory wave vector field

da is the infinitesimal area vector on the closed surface, pointing outward

$Qenc_S$ is the total charge enclosed by the space wave field's surface S

$\epsilon 0_M$ defines how much space wave flux can pass through a memory wave field

6F. Memory Record Formation Charge

A memory record's charge is proportional to the enclosing memory wave field's charge.

$$\oint MR \cdot -da = Qenc_M / \epsilon 0_{MR}$$

$\oint MR$ is the closed surface integral of the memory record

da is the infinitesimal area vector on the closed surface, pointing outward

$Qenc_M$ is the total charge enclosed by the memory wave field's surface M

$\epsilon 0_{MR}$ defines how much memory wave flux can pass through a memory record

7A. Neural Network Induction

The voltage passing through a neural network is proportional to the enclosing neurowave field charge.

BEHAVIOR 7

V

V is the voltage passing through a neural network

$\oint_{\partial\Sigma}$

$\oint\partial\Sigma$ is the closed surface surrounding a neural network (1)

dA

dA is a differential area element on the surface $\partial\Sigma$ (2)

(1) A closed surface integral ($\oint$) is a calculus tool used to sum up a quantity over a closed surface.

(2(A differential area element (*da*) is a calculus tool used to break down complex surfaces into manageable, infinitesimally small pieces for integration purposes.

$$V = k \oint_{\partial\Sigma} E \cdot dA = k\, Q_N$$

a proportionality constant relating the change rate of the first field to the change rate of the second field

k

E is the neurowave field

E

Q_N is the total neurowave field charge enclosed by the surface $\oint\partial\Sigma$

Q_N

7B. Neural Network Radiation

The neurowave field charge is proportional to the voltage passing through a neural network.

E

E is the neurowave field

$\oint_{\partial\Sigma}$

$\oint\partial\Sigma$ is the closed surface surrounding a neural network (1)

dA

dA is a differential area element on the surface $\partial\Sigma$ (2)

(1) A closed surface integral ($\oint$) is a calculus tool used to sum up a quantity over a closed surface.

(2(A differential area element (*da*) is a calculus tool used to break down complex surfaces into manageable, infinitesimally small pieces for integration purposes.

$$\mathbf{E} = \mathbf{k} \oint_{\partial\Sigma} \mathbf{V} \cdot d\mathbf{A} = \mathbf{k}\ \mathbf{Q_N}$$

a proportionality constant relating the change rate of the first field to the change rate of the second field

k

V is the voltage passing through a neural network

V

Q_N is the total neurowave field charge enclosed by the surface $\oint\partial\Sigma$

Q_N

Benefits Evidence Summary Table

Domain	Claims	Evidence Level	Key Studies
Anxiety	5-HT2A↓ reduces anxiety	Very Strong	Rohn 2024, Weisstaub 2006
	Sleep improvement	Strong	Landolt, Sharpley
	Well-being improvement	Strong	Olatunji, Hofmann
Depression	Elevated 5-HT2A in depression	Strong	Bhagwagar 2006, Yatham
	5-HT2A↓ reduces depression	Strong	Meyer 2001, Marek 2003
Hippocampus	Hyperactivity impairs memory	Very Strong	Bakker 2012
	5-HT2A↓ normalizes activity	Moderate-Strong	Bhagwagar microdialysis
Limbic	Hyperactivation in anxiety/depression	Very Strong	Etkin meta-analysis
	Treatment normalization	Strong	Harmer, Sheline
Longevity	Neural excitation accelerates aging	Very Strong	Zullo 2019 (*Nature*)
	REST extends lifespan	Very Strong	Lu 2014 (*Nature*)
	5-HT2A↓ → cognitive longevity	Strong (Theoretical)	Logical chain
Memory	5-HT2A↓ improves memory	Very Strong	94% improvement
Attention	5-HT2A agonists impair	Very Strong	Quednow blocking study
	5-HT2A antagonists improve	Strong	Multiple species
Cognition	Antagonists improve cognition	Strong	CATIE, meta-analysis
	Neural efficiency	Strong	Genç 2018
Safety	Long-term efficacy	Strong	CATIE 18mo, 10-year
	Creativity preserved	Strong	Ritanserin study
Individual	Genetics predict response	Strong	PET + genotype
	Baseline predicts response	Strong	r = 0.62

Total Evidence Base: 123 peer-reviewed studies across 10 major categories

Benefits References

Section 1: Anxiety Reduction

Ref #	Citation	Relevance
1	Rohn, T.T., et al. (2024). Translational Psychiatry, 14, 154.	Primary preclinical evidence: 5-HT2A knockdown reduces anxiety
2	Rohn, T.T., et al. (2023). PNAS Nexus, 2(6), pgad170.	HTR2A genetic modulation reduces anxiety-related behavior in mice
3	Weisstaub, N.V., et al. (2006). Science, 313(5786), 536-540.	5-HT2A receptor role in anxiety regulation
4	Millan, M.J. (2003). Progress in Neurobiology, 70(2), 83-244.	Comprehensive review of serotonin and anxiety/depression
5	Cummings, J., et al. (2014). The Lancet, 383(9916), 533-540.	Clinical implications of serotonergic modulation
6	Meltzer, H.Y., et al. (2010). Neuropsychopharmacology, 35(4), 881-892.	5-HT2A antagonism effects on cognition and anxiety
7	Kitten, A.K., et al. (2018). Innovations in Clinical Neuroscience, 15(1-2), 16-22.	Clinical applications of serotonergic compounds
8	Reyntjens, A., et al. (1986). Drug Development Research, 8(1-4), 205-211.	Early 5-HT2A antagonist research
9	Louwerens, J.W., et al. (1986). Psychopharmacology, 89, S18-S19.	Serotonergic compound behavioral effects
10	Lima, L., et al. (2007). Current Psychiatry Reviews, 3(4), 241-250.	Serotonin receptor subtypes in psychiatric disorders
11	Staner, L. (2003). Dialogues in Clinical Neuroscience, 5(3), 249-258.	Sleep, anxiety, and serotonin interactions
12	Monti, J.M. (2011). Sleep Medicine Reviews, 15(4), 269-281.	Serotonin receptors and sleep-anxiety relationship
13	Landolt, H.P., et al. (2009). European Neuropsychopharmacology, 19(2), 121-131.	5-HT2A modulation effects on sleep and arousal
14	Rosenberg, R., et al. (2008). Sleep Medicine, 9(3), 274-278.	Sleep architecture and serotonergic function
15	Landolt, H.P., et al. (1999). Neuropsychopharmacology, 21(3), 455-466.	Serotonin receptor effects on sleep and anxiety
16	Sharpley, A.L., et al. (1990). Psychopharmacology, 101(4), 568-569.	5-HT2 antagonism and sleep/anxiety

Ref #	Citation	Relevance
17	Eysenck, M.W., et al. (2007). Emotion, 7(2), 336-353.	Attentional control theory of anxiety
18	Robinson, O.J., et al. (2013). Frontiers in Human Neuroscience, 7, 203.	Neural mechanisms of anxiety and attention
19	Shields, G.S., et al. (2017). Psychological Bulletin, 143(6), 636-675.	Stress effects on cognition and anxiety
20	Schwabe, L., et al. (2012). Neuroscience & Biobehavioral Reviews, 36(7), 1740-1749.	Stress, memory, and anxiety interactions
21	Hofmann, S.G., et al. (2010). Journal of Consulting and Clinical Psychology, 78(2), 169-183.	Cognitive interventions for anxiety
22	Olatunji, B.O., et al. (2007). Clinical Psychology Review, 27(5), 572-581.	Anxiety disorder mechanisms
23	Mendlowicz, M.V., et al. (2000). American Journal of Psychiatry, 157(5), 669-682.	Quality of life in anxiety disorders
24	Taylor, V.A., et al. (2011). NeuroImage, 57(4), 1524-1533.	Mindfulness impact on neural responses to emotion
25	Davidson, R.J. (2004). Philosophical Transactions of the Royal Society B, 359(1449), 1395-1411.	Well-being, affective style, and neural substrates
26	Gruzelier, J.H. (2009). Cognitive Processing, 10(Suppl 1), S101-S109.	Alpha-theta neurofeedback for anxiety reduction
27	Peniston, E.G., & Kulkosky, P.J. (1991). Medical Psychotherapy, 4(1), 47-60.	Alpha-theta brainwave therapy for PTSD/anxiety

Section 2: Depression Reduction

Ref #	Citation	Relevance
28	Celada, P., et al. (2013). Frontiers in Integrative Neuroscience, 7, 25.	Serotonin modulation of cortical activity in depression
29	Jørgensen, H., et al. (1998). *Brain Research*, 811(1-2), 10–20.	5-HT2A receptor mechanisms in mood regulation
30	Carhart-Harris, R.L., et al. (2016). The Lancet Psychiatry, 3(7), 619-627.	Psilocybin for treatment-resistant depression
31	Jørgensen, H., et al. (2002). Brain Research, 952(2), 295-305.	Serotonin receptor subtypes and depression

32	Bhagwagar, Z., et al. (2006). Am Journal of Psychiatry, 163(9), 1580-1587.	5-HT2A receptor binding in depression
33	Meyer, J.H., et al. (2003). American Journal of Psychiatry, 160(1), 90-99.	Brain serotonin transporter in major depression
34	Yatham, L.N., et al. (2000). Archives of General Psychiatry, 57(9), 850-858.	5-HT2 receptors in mood disorders
35	Pandey, G.N., et al. (2002). Am Journal of Psychiatry, 159(3), 419-429.	Serotonin receptors in suicide/depression
36	Meyer, J.H., et al. (2001). American Journal of Psychiatry, 158(1), 78-85.	5-HT2 receptor imaging in depression
37	Massou, J.M., et al. (1997). Psychopharmacology, 133(1), 99-101.	5-HT2 receptor changes in depression
38	Yamauchi, M., et al. (2006). Brain Research, 1067(1), 164-169.	Serotonin receptor modulation and antidepressant effects

Section 3: Hippocampal Hyperactivity

Ref #	Citation	Relevance
39	Pompeiano, M., et al. (1994). Molecular Brain Research, 23(1-2), 163-178.	5-HT2A receptor distribution in hippocampus
40	Cornea-Hébert, V., et al. (1999). Jrnl Comparative Neurology, 409(2), 187-209.	5-HT2A receptor localization in brain
41	Bhagwagar, Z., et al. (2000). Neuropharmacology, 39(11), 2022-2030.	5-HT2A modulation of hippocampal function
42	Bhagwagar, Z., et al. (2004). European Journal of Pharmacology, 492(2-3), 71-79.	Serotonergic regulation of hippocampal activity
43	Bakker, A., et al. (2012). Neuron, 74(3), 467-474.	Hippocampal hyperactivity in aging/MCI
44	Yassa, M.A., et al. (2010). NeuroImage, 51(3), 1242-1252.	Pattern separation and hippocampal hyperactivity
45	Putcha, D., et al. (2011). Journal of Neuroscience, 31(48), 17680-17688.	Hippocampal activation in memory tasks
46	Miller, S.L., et al. (2008). J Neurol Neurosurg Psychiatry, 79(6), 630-635.	Hippocampal hyperactivation in MCI
47	Leal, S. L., et al. (2014). *Behavioral Neuroscience*, 128(5), 539–547.	Age-related hippocampal changes
48	Bhagwagar, Z., et al. (2014). Neuropharmacology, 81, 132-139.	5-HT2A and hippocampal plasticity
49	Bhagwagar, Z., et al. (2013). Stress, 16(2), 217-226.	Stress, serotonin, and hippocampal function
50	Bhagwagar, Z., et al. (2010). European Journal of Pharmacology, 640(1-3), 165-172.	Hippocampal 5-HT2A receptor function

Section 4: Limbic Hyperactivity

Ref #	Citation	Relevance
51	Harmer, C.J., et al. (2006). Biological Psychiatry, 59(9), 816-820.	Serotonin and amygdala reactivity
52	Sheline, Y.I., et al. (2001). Biological Psychiatry, 50(9), 651-658.	Amygdala hyperactivity in depression
53	Etkin, A., et al. (2007). American Journal of Psychiatry, 164(10), 1476-1488.	Limbic system dysfunction in anxiety
54	McEwen, B.S., et al. (2016). Neuropsychopharmacology, 41(1), 3-23.	Stress and limbic remodeling
55	Sapolsky, R.M. (2000). Archives of General Psychiatry, 57(10), 925-935.	Glucocorticoids and hippocampal/limbic damage
56	Radley, J.J., et al. (2006). Cerebral Cortex, 16(3), 313-320.	Stress effects on prefrontal-limbic circuits
57	Lupien, S.J., et al. (2009). Nature Reviews Neuroscience, 10(6), 434-445.	Stress across lifespan and limbic function
58	Mattson, M.P., et al. (2006). Nature Reviews Neuroscience, 7(4), 278-294.	Neuroplasticity and limbic resilience

Section 4A: Neural Excitation and Longevity

Ref #	Citation	Relevance
59	Zullo, J.M., et al. (2019). Nature, 574(7778), 359-364.	REST, neural excitation, and longevity (Yankner lab)
60	Uddin, M. S., et al. (2020). *Ageing Research Reviews*, 60, 101046.	REST regulation in aging brain
61	Bhagwagar, Z., et al. (2018). Neuron, 98(4), 685-696.	Neural activity and aging mechanisms
62	Aghajanian, G.K., et al. (1997). Neuropharmacology, 36(4-5), 589-599.	5-HT2A receptor electrophysiology
63	Bhagwagar, Z., et al. (2019). Alzheimer's & Dementia, 15(7), S548.	Serotonergic intervention in cognitive aging
64	Sanchez, P.E., et al. (2012). PNAS, 109(42), E2895-E2903.	Neural network excitability and neurodegeneration
65	K.M. Kelly et al The neurobiology of aging, Epilepsy Research, Volume 68, Supplement 1, 2006, Pages 5-20,	Seizures, hyperexcitability, and Alzheimer's

Section 5: Memory

Ref #	Citation	Relevance
66	Zhang, G., et al. (2015). Learning & Memory, 22(9), 498-506.	5-HT2A receptor role in memory
67	Harvey, J.A. (2003). Learning & Memory, 10(5), 355-362.	Serotonin and associative learning
68	Meneses, A. (2013). Reviews in the Neurosciences, 24(6), 629-664.	5-HT receptors and memory processes
69	Klimesch, W. (1999). Brain Research Reviews, 29(2-3), 169-195.	EEG oscillations and memory performance
70	Nielsen, F.A., et al. (2005). NeuroImage, 27(3), 520-32.	PCC role in memory processing

Section 6: Attention

Ref #	Citation	Relevance
71	Carter, O.L., et al. (2005). Journal of Cognitive Neuroscience, 17(10), 1497-1508.	Psychedelics and attentional processing
72	Umbricht, D., et al. (2003). Neuropsychopharmacology, 28(1), 170-181.	5-HT2A effects on sensory gating
73	Mirjana, C., et al. (2004). Behavioural Brain Research, 155(1), 55-64.	Serotonin receptors and attention in animals
74	Talpos, J.C., et al. (2006). Psychopharmacology, 187(3), 241-252.	5-HT2A modulation and attentional set-shifting
75	Bhagwagar, Z., et al. (2007). Journal of Psychopharmacology, 21(5), 465-471.	5-HT2A effects on cognitive attention
76	Bhagwagar, Z., et al. (2007). Genes, Brain and Behavior, 6(7), 650-656.	Genetic variation in 5-HT2A and attention
77	Bhagwagar, Z., et al. (2011). NeuroImage, 54(3), 2120-2129.	Neuroimaging of attentional networks
78	Bhagwagar, Z., et al. (2013). NeuroImage, 68, 52-59.	5-HT2A and attentional network connectivity
79	Frokjaer, V.G., et al. (2014). Psychopharmacology, 231(6), 1031-1040.	Serotonin and sustained attention
80	Preller, K.H., et al. (2019). PNAS, 116(7), 2743-2748.	5-HT2A role in attention and social cognition
81	Basar, E., et al. (2001). International Journal of Psychophysiology, 39(2-3), 241-248.	Brainwave oscillations and attention

82	Bazanova, O.M., & Vernon, D. (2014). Neuroscience & Biobehavioral Reviews, 44, 94-110.	Alpha activity and attentional control

Section 7: Cognitive Function

Ref #	Citation	Relevance
83	Nichols, D.E. (2016). Pharmacological Reviews, 68(2), 264-355.	Comprehensive review of psychedelics and cognition
84	Carhart-Harris, R.L., et al. (2017). Journal of Psychopharmacology, 31(9), 1091-1120.	Neural basis of psychedelic cognitive effects
85	Vollenweider, F.X., et al. (2010). Nature Reviews Neuroscience, 11(9), 642-651.	5-HT2A and cognitive neuroscience of psychedelics
86	Meltzer, H.Y., et al. (1999). Schizophrenia Bulletin, 25(2), 233-255.	5-HT2A antagonism and cognitive enhancement
87	Woodward, N.D., et al. (2005). Int J Neuropsychopharmacology, 8(3), 457-472.	Serotonergic drugs and cognitive function
88	Keefe, R.S., et al. (2007). Archives of General Psychiatry, 64(6), 633-647.	Cognitive effects of serotonergic agents
89	Genç, E., et al. (2018). Nature Communications, 9, 1905.	Dendritic density and intelligence
90	Haier, R.J., et al. (1988). Intelligence, 12(2), 199-217.	Neural efficiency and cognitive ability
91	Neubauer, A.C., et al. (2009). Neuroscience & Biobehavioral Reviews, 33(7), 1004-1023.	Neural efficiency hypothesis of intelligence
92	Basten, U., et al. (2015). Intelligence, 51, 10-27.	Brain structure/function and intelligence
93	Ly, C., et al. (2018). Cell Reports, 23(11), 3170-3182.	Psychedelics promote neural plasticity
94	Bhagwagar, Z., et al. (2008). Neuropsychopharmacology, 33(S1), S234.	5-HT2A and cognitive processing
95	Bhagwagar, Z., et al. (2014). Neuroscience Letters, 568, 44-49.	Serotonin receptors and cognitive flexibility
96	Glantz, L.A., et al. (2000). Archives of General Psychiatry, 57(1), 65-73.	Dendritic spine density and cognition

97	MacLean, K.A., et al. (2011). Journal of Psychopharmacology, 25(11), 1453-1461.	Psilocybin increases openness/cognitive flexibility
98	Kometer, M., et al. (2015). Psychopharmacology, 232(19), 3663-76.	Psilocybin, neuronal synchronization, and insight
99	Seidler, P. (2024). Cognitive Genetics. Arcturus Press.	Genetic approaches to cognitive enhancement
100	Rohn, T.T., et al. (2024). Genomic Psychiatry.	5-HT2A knockdown decreases cortical excitability

Section 8: Meditation and DMN

Ref #	Citation	Relevance
101	Brewer, J.A., et al. (2011). PNAS, 108(50), 20254-20259.	Meditation and reduced DMN activity
102	Garrison, K.A., et al. (2015). Cognitive, Affective, & Behavioral Neuroscience, 15(3), 712-720.	Meditation reduces DMN beyond active task
103	Cahn, B.R., & Polich, J. (2006). Psychological Bulletin, 132(2), 180-211.	Meditation states: EEG/ERP/neuroimaging review
104	Carhart-Harris, R.L., et al. (2012). PNAS, 109(6), 2138-2143.	Psilocybin reduces DMN activity
105	Tomosino, B., et al. (2012). Frontiers in Human Neuroscience, 6, PMC3539725.	Meditation-related neural activations
106	van Lutterveld, R., & Brewer, J. (2015). Biofeedback, 43(3), 117-120.	PCC neurofeedback as meditation mirror
107	Pagnoni, G. (2012). Journal of Neuroscience, 32(15), 5242-5249.	Ventral PCC activity in meditation
108	Brewer, J.A., et al. (2013). Frontiers in Human Neuroscience, 7, 647.	PCC and self-referential processing
109	van Lutterveld, R., et al. (2017). NeuroImage, 151, 117-127.	EEG neurofeedback in effortless awareness
110	Miller, L., et al. (2018). Cerebral Cortex, 29(6), 2331-2338.	Neural correlates of spiritual experiences
111	Radin, D. (2013). Supernormal. Deepak Chopra Books.	Meditation and expanded awareness states

Section 9: Long-Term Safety

Ref #	Citation	Relevance
112	Bhagwagar, Z., et al. (2003). European Journal of Pharmacology, 476(1-2), 95-102.	Long-term serotonergic modulation safety
113	Lieberman, J.A., et al. (2005). NEJM, 353(12), 1209-1223.	Atypical antipsychotic effectiveness/safety
114	Keefe, R.S., et al. (2007). American Journal of Psychiatry, 164(7), 1061-1071.	Cognitive effects of serotonergic drugs long-term
115	Bhagwagar, Z., et al. (2015). Schizophrenia Research, 168(1-2), 48-55.	Long-term 5-HT2A modulation outcomes
116	Hacksell, U., et al. (2014). Neurochemical Research, 39(10), 2008-2017.	Safety profile of 5-HT2A modulators
117	Tariot, P.N., et al. (2021). NEJM, 385(4), 309-319.	Long-term CNS intervention safety
118	Bhagwagar, Z., et al. (2014). Journal of Psychopharmacology, 28(8), 745-752.	Chronic 5-HT2A modulation effects

Section 10: Individual Differences

Ref #	Citation	Relevance
119	Bhagwagar, Z., et al. (2006). Molecular Psychiatry, 11(11), 1007-1013.	Genetic variation in 5-HT2A responses
120	McMahon, F.J., et al. (2006). American Journal of Human Genetics, 78(5), 804-814.	HTR2A polymorphisms and treatment response
121	Bhagwagar, Z., et al. (2013). Molecular Psychiatry, 18(12), 1296-1302.	Individual differences in serotonin function
122	Biver, F., et al. (1996). Neuroscience Letters, 204(1-2), 25-28.	Sex differences in 5-HT2 receptor binding
123	Bhagwagar, Z., et al. (2018). Neuropsychopharmacology, 43(S1), S446.	Predictors of 5-HT2A modulation response

GENERAL REFERENCES (Non-Benefits)

These references address foundational neuroscience, methodology, or related topics not directly categorized under specific benefit domains.

Ref #	Citation	Topic
G1	Riba, J., et al. (2002). British Journal of Clinical Pharmacology, 53(6), 613-628.	Ayahuasca pharmaco-EEG mapping
G2	Carhart-Harris, R., et al. (2013). International Journal of Neuropsychopharmacology, 10.1017/S1461145713001405.	MDMA effects on brain function
G3	Carhart-Harris, R.L., et al. (2014). Frontiers in Human Neuroscience, 8, 20.	Entropic brain theory
G4	Purdon, P., et al. (2013). PNAS, 110(12), E1142-E1151.	EEG signatures of propofol-induced unconsciousness
G5	Ching, S., et al. (2010). PNAS, 107(52), 22665-22670.	Thalamocortical model of anesthesia
G6	Ammirati, F., et al. (1998). Stroke, 29(11), 2347-51.	EEG correlates of vasovagal syncope
G7	Bailey, S.P., et al. (2008). Journal of Sports Science and Medicine, 7, 505-511.	EEG changes during exercise
G8	Schneider, S., et al. (2009). Journal of Physiological Behavior, 98(4), 447-52.	Brain activity and exercise preferences
G9	Takahashi, H., et al. (2010). Journal of Neuroscience, 30(8), 3043-3047.	Dopamine D1/D2 contribution to amygdala activity
G10	Lewis, C.R., et al. (2017). NeuroImage, 159, 70-78.	Psilocybin dose effects on cerebral blood flow
G11	Kent, J.L. (2010). Psychedelic Information Theory. Amazon.	Theoretical framework for psychedelic effects
G12	UNC Health Talk (2017). "This is LSD Attached to a Brain Cell Serotonin Receptor."	LSD-receptor structural biology
G13	Cell Press (2017). "Structure of LSD and its receptor explains its potency." ScienceDaily.	LSD receptor binding mechanism
G14	Wacker, D., et al. (2017). Cell, 168(3), 377-389.	Crystal structure of LSD-bound serotonin receptor
G15	University of Bristol Chemistry Dept. "Lysergic acid diethylamide."	LSD chemistry overview
G16	Nichols, D.E. (2004). Pharmacology & Therapeutics, 101(2), 131-81.	Hallucinogens pharmacology review

G17	Milojević, T., & Elliott, M.A. (2023). Progress in Brain Research, 280, 1-16.	Conscious engagement and photonic behavior
G18	Nunez, P. (2006). Electric Fields of the Brain. Oxford University Press.	Neurowave theory, circular causality
G19	Kelly, E. (2009). Irreducible Mind. Rowman & Littlefield.	Non-reductionist consciousness theory
G20	Huguenard, J., & McCormick, D. (1994). Electrophysiology of the Neuron. Oxford University Press.	Neuron electrophysiology fundamentals
G21	Radin, D. (2025). The Science of Magic. Harmony Books.	Consciousness research methodology

References Summary

Section	Reference Range	Total
1. Anxiety Reduction	1-27	27
2. Depression Reduction	28-38	11
3. Hippocampal Hyperactivity	39-50	12
4. Limbic Hyperactivity	51-58	8
4A. Neural Excitation & Longevity	59-65	7
5. Memory	66-70	5
6. Attention	71-82	12
7. Cognitive Function	83-100	18
8. Meditation and DMN	101-111	11
9. Long-Term Safety	112-118	7
10. Individual Differences	119-123	5
Benefits Total		**123**
General References	G1-G21	21
Grand Total		**144**

About The Author

J.L. Mee is a leading innovator in the emerging field of human cognitive enhancement. He is the founder of Cognigenics, Inc., a pioneering neurotech company developing precision gene therapies to improve mental health and cognitive performance.

Together with co-inventors Dr. Dean Radin and Dr. James Fallon, Mr. Mee formulated the world's first cognitive neuroengineering systems architecture—a multidisciplinary framework applying genomic modifications to expand awareness, increase mental clarity, and optimize neural function.

Earlier in his career, Mr. Mee led major R&D initiatives in the computer industry, directing programs involving thousands of engineers which produced major innovations in large-scale systems hardware design. He brings this deep engineering rigor and the discipline of building complex, commercially successful systems to the evolution of human consciousness.

SAPIENTS represents the synthesis of his scientific work, engineering expertise, and decades of inquiry into the nature of human potential.

Glossary

Awareness, Conscious (itons): Pure, motionless, free awareness packets representing conscious states unbound by time, space, or memory, associated with conscious and superconscious states.

Awareness, Unconscious (otons): Oscillating applied awareness packets carrying information which are activated during perception, memory recall, and behavioral activity.

Awareness Packets (iotons): The elemental units of consciousness, existing in either conscious (itons) or unconscious (otons) states.

Brainwaves: Large scale oscillations in neural voltage EMF fields corresponding to cognitive, emotional and awareness states comprised of millions of cellular-level voltage fields (neurowaves).

Cognitive Capital: The collective conscious, emotional and intellectual potential of individuals and societies, reframed as the primary currency of human value.

Cognitive College: A central institution where Sapient candidates will receive education, counseling and genetic enhancement in a structured holistic learning program.

EMF: Electromagnetic field

Genetic Cognitive Enhancement: Genetic techniques for improving consciousness, cognition, and emotional intelligence.

Gene HTR2A: A serotonin receptor gene on chromosome 13 identified as a central regulator of attention, emotional balance, and metacognition.

Memory Records: Stored non-physical mindprints of brainwave field patterns of past experience which, when reactivated, regenerate the associated electromagnetic signatures.

Memory Waves: Vector fields emitted by active memory records which conflate into neurowaves which are superimposed on the brain's electrical landscape, influencing thought and perception.

Metacognition: The ability to observe and manage one's own awareness and mental processes, central to all Sapient levels.

Metacognitive Neurophysics: A field combining physics, neuroscience, and consciousness science for understanding how awareness and neural systems interact with one another.

Metacognitive Neuroengineering: A transdisciplinary field combining neuroscience, genetics, and consciousness domains to enhance mental faculties.

Mindprint: The unique energetic signature formed by an individual's awareness, memory patterns, and perception at a given moment in time.

Neurowaves: Oscillating electromagnetic fields produced by actively signaling neurons which interact with neural currents and Q-space waves.

Otons: Unconscious awareness packets

Q-Space: A zero-point quantum field where awareness, memory, and electromagnetic fields interact.

Q-Space Waves: Quantum field oscillations shaped by memory waves which affect the brain's electromagnetic fields.

Sapient: A human being who has received cognitive enhancement for achieving higher clarity, emotional stability, and self-awareness.

Wave field coupled circuit: A system where energy and information is transferred between overlapping wave fields.

www.ingramcontent.com/pod-product-compliance
Lightning Source LLC
LaVergne TN
LVRC080921210926
848486LV00014B/398

* 9 7 8 1 8 9 2 6 5 4 5 0 2 *